Frontiers in Molecular Biology

Series editors

B.D.Hames
Department of Biochemistry, University of Leeds, Leeds LS2 9JT, UK

D.M.Glover
Cancer Research Campaign, Eukaryotic Molecular Genetics Research Group, Department of Biochemistry, Imperial College of Science and Technology, London SW7 2AZ, UK

Other titles in the series:

Gene Rearrangement

Genes and Embryos

Molecular Immunology

Oncogenes

Transcription and Splicing

MOLECULAR NEUROBIOLOGY

Edited by

D.M.Glover

Cancer Research Campaign, Eukaryotic Molecular Genetics
Research Group, Department of Biochemistry, Imperial College
of Science and Technology, London SW7 2AZ, UK

and

B.D.Hames

Department of Biochemistry, University of Leeds, Leeds LS2 9JT, UK

IRL PRESS
at
OXFORD UNIVERSITY PRESS
Oxford New York Tokyo

Oxford University Press
Walton Street, Oxford OX2 6DP

Oxford is a trade mark of Oxford University Press

Published in the United States
by Oxford University Press, New York

© Oxford University Press, 1989

*All rights reserved. No part of this publication may be reproduced,
stored in a retrieval system, or transmitted, in any form or by any means,
electronic, mechanical, photocopying, recording, or otherwise, without
the prior permission of Oxford University Press*

*This book is sold subject to the condition that it shall not, by way
of trade or otherwise, be lent, re-sold, hired out, or otherwise circulated
without the publisher's prior consent in any form of binding or cover
other than that in which it is published and without a similar condition
including this condition being imposed on the subsequent purchaser*

British Library Cataloguing in Publication Data

Molecular neurobiology
1. Neurobiology. Molecular Biology.
I. Hames, B.D. (B.David). II. Glover,
David M. III. Series.
591.1'88

ISBN 0 19 963042 9
ISBN 0 19 963043 7 pbk.

Library of Congress Cataloging in Publication Data
(Data available)

ISBN 0 19 963042 9
ISBN 0 19 963043 7 (Pbk)

Previously announced as:
ISBN 1 85221 082 6
ISBN 1 85221 078 8 (Pbk)

Typeset and printed by Information Press Ltd, Oxford, England.

Preface

This book examines a number of areas in which molecular approaches are being used in neurobiology. We have made no attempt to be comprehensive, because of the vast topic area. Rather we have sought to highlight three areas which lend themselves to molecular studies, and in which significant advances have been made. Toni Claudio describes work on the molecular characterization of ion channels, a field that has benefited greatly from DNA cloning techniques. She has concentrated upon the characterization of the acetylcholine receptor, since this system has been most intensively studied, and it exemplifies approaches that are now being used to characterize other channels. A genetic approach has much to offer in studies of neurophysiology. This is an area in which *Drosophila melanogaster* comes into its own as an experimental organism. It is but a short step from *Drosophila* genes to their mammalian equivalents, as seen in the cloning of the potassium channel gene first from flies and subsequently from mammals using *Drosophila* gene probes. *Drosophila* genetics may be somewhat of a mystery area for neurobiologists and other molecular biologists alike. We feel sure therefore that the chapter by Ganetzky and Wu will be both illuminative and fascinating for its readers. Finally Jon Covault looks at the molecular biology of cell adhesion in neural development, problems highly suited for molecular studies. Our thanks go to each of these authors for their patience in this project, and also to Eric Barnard for setting the scene for the book with his overview of the field. As they all know, we would have liked to have put so much more into this small volume, but nevertheless we hope it will provide a useful window on some topical areas.

David Hames
David Glover

Contributors

Eric A.Bernard
MRC Molecular Neurobiology Unit, Hills Road, Cambridge, UK

Toni Claudio
Department of Cellular and Molecular Physiology, Yale University School of Medicine, 333 Cedar Drive, New Haven, CT 06510, USA

Jon Covault
University of Connecticut, Department of Physiology and Neurobiology, Storrs, CT 06269, USA

Barry Ganetzky
Laboratory of Genetics, University of Wisconsin, Madison, WI 53706, USA

Chun-Fang Wu
Department of Biology, University of Iowa, Iowa City, IA 52242, USA

Contents

Abbreviations	xi
1. Molecular neurobiology—an introduction E.A.Barnard	1
2. Molecular approaches to neurophysiology in *Drosophila* B.Ganetzky and C.-F.Wu	9
Introduction	9
Scope	9
Electrical signaling in the nervous system	9
Electrophysiological methods in *Drosophila*	12
Strategies for cloning genes in *Drosophila*	14
Isolation of mutants	17
Germ line transformation and *in-vitro* mutagenesis	18
Molecular and Genetic Analysis of the Cholinergic System	18
Cha: the gene for choline acetyltransferase	19
Ace: the gene for acetylcholine esterase	20
A gene for an acetylcholine receptor subunit	22
Molecular and Genetic Analysis of Phototransduction	23
The R1–R6 opsin gene: *ninaE*	26
Genes for minor forms of opsin	27
Genes affecting intermediate steps of phototransduction	29
Molecular and Genetic Analysis of Sodium Channels	31
A *Drosophila* homolog of the vertebrate sodium channel	31
Mutations affecting sodium channels	34
Double mutant interactions	41
Molecular and Genetic Analysis of Potassium Channels	43
Mutations affecting potassium channels	45
Interaction between sodium and potassium channel mutations	53
Conclusions	54
Acknowledgements	54
References	55

3. Molecular genetics of acetylcholine receptor-channels 63
T. Claudio

Introduction	63
Ligand-gated receptor-channels	64
Voltage-gated channels	66
Background to the AChR	67
Composition and structure	68
Pharmacology	69
Physiology	69
Development	71
Immunology	73
Isolation of *Torpedo* AChR cDNAs	73
Structural Predictions Based on Deduced Amino Acid Sequences from *Torpedo* AChR cDNAs	76
Size of each subunit	76
Potential glycosylation sites	78
Potential phosphorylation sites	80
Location of the ACh binding site	80
Topology of folding of subunits in the plane of the membrane	81
Isolation of Nicotinic AChR cDNAs From Other Species	87
Isolation of AChR cDNAs	90
Sequence similarities	90
Location of channel-lining sequences	92
Isolation of Novel AChR Subunits	95
An ϵ subunit	96
Two *Xenopus* α subunits	97
Expression Systems	98
In-vitro translation systems	98
Escherichia coli	98
Xenopus oocytes	99
Yeast	100
Stable expression in cultured cell lines	100
Determining the Function of Individual AChR Subunits	102
Different combinations of AChR subunits	102
Hybrid and chimeric AChRs	104
Configuration of subunits around the channel	107
Speculations on the order of subunit assembly	109
Isolation of Chromosomal Genes for Nicotinic AChRs	112
Genomic organization	112
Transcription of AChR genes	113
Isolation of Neuronal AChRs	114
Other Cloned Channels	116
Cloned ligand-gated receptor-channels	116
Cloned voltage-gated channels	120

Conclusions	124
Acknowledgements	126
References	126

4. Molecular biology of cell adhesion in neural development   **143**
J. Covault

Introduction	143
A Diversity of Cell Adhesion Molecules	147
NCAM—the first 'neural cell adhesion molecule'	147
L1/NgCAM—a second calcium-independent neural adhesion molecule	151
Myelin-associated glycoproteins	153
NCAM, L1/NgCAM, MAG, and P_0 have homologous extracellular domains	154
Calcium-dependent neural adhesion molecules	157
J1/cytotactin and the L2/HNK-1 family of cell adhesion molecules	158
AMOG—adhesion molecule on glia	160
Extracellular matrix adhesion molecules	161
Involvement of Cell Adhesion Molecules in Neural Development	166
Modulation of cell–cell adhesion during primary neural induction	167
Neural crest cell migration	168
Cerebellar granule cell migration	170
Axon growth	175
Cell adhesion molecules and neural specificity	181
Nerve–muscle adhesion	184
Conclusions	188
Acknowledgements	189
References	189

Index   **201**

Abbreviations

ACh	acetylcholine
AChR	acetylcholine receptor
BuTx	bungarotoxin
CNS	central nervous system
DHP	1,4-dihydropyridine
DLM	dorsal longitudinal muscle
EMS	ethylmethane sulfonate
ERG	electroretinogram
GABA	γ-aminobutyric acid
GFP	giant fiber pathway
PNS	peripheral nervous system
PSI	peripherally synapsing interneuron
STX	saxitoxin
TTX	tetrodotoxin

1

Molecular neurobiology— an introduction

Eric A.Barnard

'From whatever place I write, you will expect that a part of my travels will consist of excursions in my own mind'.

Samuel Taylor Coleridge, 1772–1834

I believe that the great majority of biological scientists in all disciplines would agree that the most revolutionary and productive development they have seen this century has been that of the new field of molecular biology. Its influence on biological thought is comparable to that of the greatest previous transformations, for example that of Darwinian evolution and the subsequent development of classical genetics and chromosome theory. The effects of molecular biology on biological science, when viewed as a whole, lie partly in its own tremendous conceptual content—the cellular mechanisms of inheritance and the encoding and read-out of structural and functional information at the molecular level. There is also another aspect of the discipline of molecular biology which is of major importance in biology as a whole, namely its creation of a new, DNA-based technology.

The ability to manipulate and fabricate DNA segments and to subsequently propagate these recombinant molecules has rapidly become an immensely exploitable tool, the fundamental significance of which is quite independent of the framework on which it is based. Specific examples are the isolation and modification of individual genes and their reintroduction into whole organisms as, for example, in the introduction of desirable and permanent characters into plant species; the use of cloned genes as probes to reveal the loci of inherited chromosomal defects in human pedigrees; the use of radiolabeled probes in *in-situ* hybridization experiments to localize mRNA populations and so trace developmental pathways in an embryo; the expression of cloned genes to produce rare mammalian proteins on a commercial scale in either bacterial or animal cell systems. The list will be endless and readers of this book will no doubt recognize many other applications of the new DNA-based technology.

Molecular neurobiology has developed from the collision of this DNA-based technology with neuroscience; itself the most important unknown territory in biological science. The enquiries of neuroscience lead to awe-inspiring objectives. How does a nervous system control function? What are the molecular substrates of behavior? What is the molecular basis of memory, learning, consciousness, emotion, cognition? What goes wrong in psychiatric disorders? Modern neurobiology has the most ambitious and hubristic program of any field of science—even, most of us would agree, beyond the profound mysteries of astrophysics. We have set out, no matter how cautiously in the beginning, upon the ultimate quest; to answer the Delphic injunction: Know thyself.

Molecular neurobiology does indeed start at the beginning in this long quest. In this beginning, we can identify a variety of macromolecules from more conventional neurobiological studies. These are particularly exemplified by the receptors for neurotransmitters, which are specialized for roles in the nervous system. Those which we can assign on present knowledge are listed in *Table 1*, but this list is growing all the time and new categories and subcategories will need to be added as recombinant DNA technology discovers them in neuronal DNA libraries. A major activity in current molecular neurobiology is, therefore, to establish the primary structure of representatives of all categories of the neuro-functional proteins seen in *Table 1*. From comparisons of structures, classifications of families and superfamilies (1) emerge which in themselves give new insights into possible roles of related molecules. Proposals on function are idle speculation until enough sequences are known.

DNA cloning in this phase is concerned with generating the tools required to exploit molecular studies within the nervous system itself. Once the gene for any component has been cloned, the agenda for molecular neurobiology should then include some or all of the following applications.

(i) Expression of all the cloned cDNAs required for the biological activity of the component concerned, either via *in-vitro* transcribed RNAs and their translation following injection of the RNAs into *Xenopus* oocytes (2), or via transfection of the DNAs into a mammalian cell culture expression system. Biological activity of expressed products can be tested, for example, by binding of specific radioligands, or electrophysiologically (for receptors and ion channels), or by antibody recognition.

(ii) Implantation, by transfection of all the DNAs required to encode a component, into permanent cell lines to express the protein of interest. Such cell lines can be useful tools for extensive or subtype-specific pharmacological studies. These can be applied to characterize receptors or ion channel types, or the lines may be used for more rational drug screening than hitherto possible.

(iii) Directed mutagenesis of the gene encoding the protein, or the generation of genes encoding hybrid or chimeric molecules. Novel proteins, engineered in this way, can be used to test hypotheses on

Table 1. Protein types essential for the specialized functions of the nervous system

Receptors for neurotransmitters and hormones
Ion channel proteins and associated signal transduction components
Brain-specific protein kinases
Precursor proteins for neuropeptide modulators
Enzymes for transmitter synthesis
Neuropeptide processing enzymes
Transmitter release machinery
Transmitter re-uptake carriers
Neuronal cytoskeleton and axonal transport proteins
Neuron-specific growth or survival trophic factors and their receptors
Neuron growth-associated proteins
Inhibitors of neuronal growth
Glial-specific growth factors and their receptors
Synaptic and dendritic morphology determinants
Neuronal cell adhesion and cell address macromolecules
Morphogens of the nervous system (other than those included above)
Proteins associated with memory mechanisms (as yet hypothetical)

the structural requirements for each of their proposed functions. Functional tests carried out on engineered proteins must be complemented by chemical studies on the proteins themselves, for example affinity labeling of ligand sites, cross-linking to test for neuropeptide binding, channel blocking, etc.

(iv) Identification in the expression systems of the requirements for signal transduction where this involves second messenger systems, and for regulation by brain protein kinases.

(v) Determination of the organization of genes for structural components, the relating of exons to functional domains and the search for common neuronal or type-specific elements in the untranslated sequences.

(vi) Removal or addition or transposition of upstream, or other, regulatory sequences in an implanted DNA sequence in cell lines, to identify signals for the regulation of expression, an approach that will be invaluable to understanding selective expression in neurons of a particular type or location.

(vii) The use of DNA probes specific for each polypeptide in a functional unit and their families, to map by *in-situ* hybridization the spatial distribution in brain and developmental time of expression of their respective mRNAs, and the control thereof by regulatory factors.

(viii) The identification, by the mRNA or antigen distributions, of proteins occurring in specific gradients during the central nervous system (CNS) development, or acting in morphogen target expression, or in other features of neuronal and glial cell recognition.

(ix) The identification of proteins undergoing changed patterns of synthesis or covalent modification in experimental paradigms of simple behavior, of learning and of memory, especially in simpler nervous systems.

(x) The use of DNA probes in screening human pedigrees or animal models (where appropriate) to identify candidate genes or lineage markers for inherited forms of neurological or psychiatric disorders.

This list does not exhaust the avenues of investigation opened up by molecular cloning, but indicates some of the routes that have already been embarked upon.

The present volume presents expert accounts of three areas where rapid and impressive progress is now being made in the applications of molecular neurobiology. The choice is eclectic: the limitations of space prevent a number of studies in other currently active fields being covered. The aim of the editors has been to provide a representative and illuminating sample of these studies.

The chapter by Ganetsky and Wu on molecular approaches to neurobiology in *Drosophila* serves to illustrate well some important methods in molecular neurobiology as a whole. The use of a far simpler organism than a mammal is of great experimental advantage in permitting an integrated application of molecular and traditional genetics. Yet the structures of the component parts, where known, have turned out to be homologous enough to their mammalian equivalents to be applied to analysis of the latter. This principle of the conservation of important structures, evidence of basically similar molecular mechanisms in most elements of the nervous system, has proven to be widely obeyed and will be of great practical value.

The chapter by Claudio on the acetylcholine (ACh) receptor shows how this receptor has been, so far, the system *par excellence* for the application of molecular biology to the analysis of a complete functioning element of the nervous system. In general, the receptors and ion channels of the nervous system offer us the best targets for analysis in this, the first stage of molecular neurobiology. Their genes are now susceptible to isolation and to cDNA cloning and they clearly represent important and clearly defined functional elements in neuronal signaling. To put the nicotinic receptor into perspective, it must be remembered that this is in a minority, though highly important, class of receptors in the nervous system. The receptor types so far known in vertebrate neurotransmission are summarized in *Table 2*. There are only ten generally agreed small-molecule neurotransmitters, although these operate a much larger number of receptors due to the multiple types and the many subtypes of the latter. A few other candidate molecules have been tentatively discussed as neurotransmitters. With peptides, on the other hand, the structural possibilities are countless and an unknown but a very large number of these are likely to act in the brain. In general, they appear to be co-transmitters or modulators of the action or release of a classical neurotransmitter.

Table 2. Known neurotransmitter and neuromodulator receptors

Class 1: *Ligand-gated ion channels*
    acetylcholine (nicotinic)
    ATP[a]
    $GABA_A$
    Glycine
    Glutamate[b]: kainate-type
                      quisqualate-type
                      NMDA-type
    $5\text{-}HT_3$

Class 2: *G-protein-coupled receptors* (1 subunit, 7 membrane-domain proteins)
    acetylcholine (muscarinic)
    adenosine
    α-adrenergic
    β-adrenergic
    dopamine
    histamine
    $5\text{-}HT_1$ and $5\text{-}HT_2$
    many neuropeptides

Class 3: *Growth-factor-type neuropeptide receptors*
  (a) Receptors for growth factors active in the nervous system [e.g. nerve growth factor (3)]
  (b) Receptors for some other peptides (4) which occur both with a neuronal function and elsewhere with other functions; their receptors (as known from their non-neuronal sites) are of the same general structure (4) as Class 3a, i.e. with a single membrane domain, e.g. the brain insulin receptor (5)

Class 4: *Guanylate cyclase peptide receptors*
    e.g. atrial natriuretic peptide (ANP) receptor

[a]Abbreviations: GABA, γ-aminobutyric acid; 5-HT, 5-hydroxytryptamine; NMDA, *N*-methyl-D-aspartate. Only vertebrate types are included here.
[b]Some glutamate receptors can also activate a second messenger system.

    The nicotinic receptor is seen to belong to a superfamily of ligand-gated ion channels (Class 1) (*Table 2*), that is receptors which contain, in the same protein molecule, their own ion channel operated allosterically by the agonist ligand. Molecular neurobiology has so far deciphered structures of the first three types of Class 1, all of which are hetero-oligomers which have some discernable sequence homology in common and four hydrophobic putative transmembrane domains in each subunit (1).
    A second superfamily of receptors has likewise been revealed by molecular neurobiology, termed Class 2 (*Table 2*). These are coupled to their effector system (e.g. via either adenylate cyclase, phosphoinositide, diacylglycerol, arachidonic acid metabolites, or directly to an ion channel) by a guanine nucleotide-binding protein (G-protein). DNA sequences have, at the time of writing, been obtained for a number of the types of Class

2 receptors. Each has a single subunit, although homo-oligomers of these may often be the functional form. All show a similar hydrophobicity pattern, with seven hydrophobic domains (predicted to be transmembrane) in each subunit. It is believed that a variety of neuropeptide receptors, as yet not analyzed, will fall into this class. Each of the known receptors in Class 2 comprises multiple subtypes with different pharmacologies and distributions, for example D1 and D2 dopamine receptors, 5-hydroxy-tryptamine (5-HT)$_{1a,b,c}$, etc.

There are at least two other classes of receptors for peptides which have a specific action of neurons, as indicated in *Table 2*. A newly discovered case is that of atrial natriuretic peptide (ANP), whose receptor has been found by DNA cloning to be a guanylate cyclase which is activated allosterically by the binding of the peptide (6). A family of peptide-activated membrane guanylate cyclases may therefore exist.

It can be seen, therefore, that the nicotinic ACh receptor is typical of one limited, but very important, class of receptors (Class 1). Numerically, however, most receptors in the nervous system appear to belong to Class 2. Molecular neurobiology has disclosed the existence and nature of these classes and it is clear that its tools—as outlined above—are capable of characterizing and exploiting each member of each class of receptors as their encoding DNAs become cloned. This will in time become true, one can be sure, for all of the other specific brain proteins recognized (*Table 1*). Completely new categories of these proteins will surely be found by further cloning expeditions and, once the component protein building blocks are identified and understood, molecular neurobiology will be applicable at the cellular level (as is clearly demonstrated in Chapter 4 of this volume by J.Covault) for the analysis of neuronal and glial interactions and the formation of neural circuits.

In the subject-matter reviewed by this volume, we are at the start of an immensely powerful enquiry into the brain and its mysteries. We have reason to hope that this discipline will lead us in its far distant development to a precise answer, in neuroanatomical and molecular terms, to Shakespeare's eternal question: 'Tell me where is fancy bred? Or in the heart or in the head?'

References

1. Barnard,E.A., Darlison,M.G. and Seeburg,P.H. (1987) Molecular biology of the GABA$_A$ receptor: the receptor/channel super-family. *Trends Neurosci.*, **10**, 502–509.
2. Barnard,E.A. and Bilbe,G. (1987) Functional expression in the *Xenopus* oocyte of mRNAs for receptors and ion channels. In *Neurochemistry: A Practical Approach*. A.J.Turner and H.Bachelard, (eds) IRL Press, Oxford, pp. 243–270.
3. Radeke,M.J., Misto,T.P., Hsu,C., Herzenberg,L.A. and Shooter,E.M. (1987) Gene transfer and molecular cloning of the rat nerve growth factor receptor. *Nature*, **325**, 593–597.

4. Iversen,L.L. (1984) Amino acids and peptides: fast and slow chemical signals in the nervous system. *Proc. R. Soc., Lond. B*, **221**, 245–260.
5. Heidenreich,K.A., de Vellis,G. and Gilmore,P.R. (1988) Functional properties of the subtype of insulin receptor found on neurons. *J. Neurochem.*, **51**, 878–887.
6. Chinkers,M., Garbers,D.L., Chang,M-S., Lowe,D.G., Chin,H., Goeddel,D.V. and Schulz,S. (1989) A membrane form of guanylate cyclase is an atrial natriuretic peptide receptor. *Nature*, **338**, 78–83.

2

Molecular approaches to neurophysiology in *Drosophila*

Barry Ganetzky and Chun-Fang Wu

1. Introduction

1.1 Scope

This article is primarily concerned with the use of classical and molecular genetic approaches for studying the macromolecules involved in electric signaling in the nervous system. One objective is to present the current state of knowledge about the various genes and gene products that mediate neuronal excitation in *Drosophila*. Another objective is to use these specific examples to illustrate the variety of approaches and techniques available in *Drosophila* that provide unique experimental opportunities in this organism for the molecular analysis of nervous system function.

1.2 Electrical signaling in the nervous system

For readers who are not familiar with the basic details of the signaling mechanisms in nervous systems we provide a brief overview to put the work that will be described into an appropriate context.

The nervous system of an organism functions to receive information about its external and internal environment, to process this information, and to produce an appropriate response. The signaling functions of nerve cells, or neurons, depend on the properties of their electrically excitable outer membrane. Neurons maintain a potential difference (resting potential) across the membrane with the inside of the cell negative relative to the outside. The resting neuron also maintains concentration gradients of various ions across the membrane. In particular, the concentration of sodium and calcium ions are relatively high outside the cell while the concentration of potassium ions is high inside the cell. Signaling within the nervous system generally involves a change in the resting membrane potential brought about by charge transfers carried by ionic fluxes through gated pores formed by transmembrane proteins called channels (*Figure*

1). When activated, ion channels selectively allow certain species of ions to pass through the membrane, driven by their concentration gradient. The membrane potential will be made either more positive or more negative depending on the direction of fluxes of the particular ion species.

Environmental stimuli are perceived by specialized neurons called sensory cells. Each type of sensory cell is specially adapted to respond

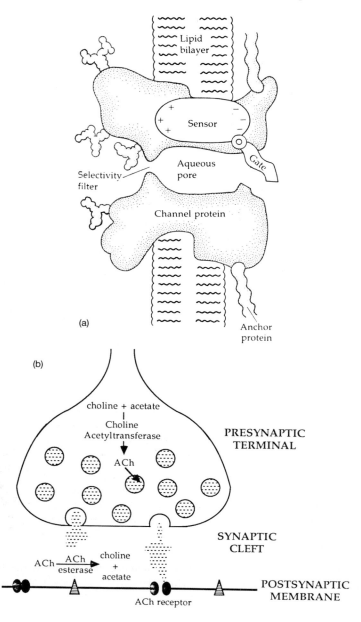

to a particular type of stimulus including light, sound, touch, heat, and certain chemicals. These cells are capable of sensory transduction, that is, the sensory energy provided by an appropriate stimulus is transformed and amplified to generate electrical neural signals. Many of the molecular details of sensory transduction are still poorly understood but ultimately the stimulus results in an alteration of the ionic permeability of the sensory cell membrane causing (usually) a depolarization of the sensory cell from its resting level. This departure from the resting potential in response to a stimulus is called a receptor potential. Generally, the amplitude and duration of the receptor potential increases logarithmically with the intensity of the stimulus. A receptor potential is a local signal that is not propagated along the neuronal process, or axon. However, the receptor potential acts as an electrical stimulus to the axonal membrane and, if the depolarization reaches a certain threshold level, triggers the production of an action potential in the axon. Action potentials are stereotypic all-or-nothing electrical impulses, with amplitudes in the order of 100 mV, that propagate themselves without distortion or attentuation along the entire length of an axon.

The ionic events underlying the generation and propagation of an action potential are as follows. When axonal membranes are depolarized, sodium channels open and allow sodium ions to rush down their electrochemical gradient into the cell producing the rapid rising (depolarizing) phase of an action potential. Within milliseconds after opening, sodium channels are inactivated, shutting off the influx of sodium ions. At about the same time, the membrane depolarization activates potassium channels and permits the efflux of potassium ions to repolarize the cell and restore the membrane resting potential. During the course of an action potential, the sodium currents flowing in one region of axonal membrane cause the depolarization above threshold of an adjacent region of membrane so that it fires an action potential in turn. This sequence of events is

Figure 1. (a) A hypothetical view of a voltage-gated ion channel. The channel forms an aqueous-filled pore across the outer cell membrane through which appropriate ions can diffuse when the channel is activated. The outer surface of the channel is modified post-translationally by glycosylation. The important functional regions of the channel including the selectivity filter, gate, and sensor are shown. Anchor proteins that are presumed to play a role in the membrane localization of channels are also indicated. Reproduced with permission from ref. 98. (b) A diagrammatic view of a cholinergic synapse. The transmitter, acetylcholine, is synthesized in the pre-synaptic terminal by choline acetyltransferase and packaged in membrane-bound vesicles. Transmitter is released in response to depolarization of the terminal and diffuses across the synaptic cleft to interact with receptor molecules on the surface of the post-synaptic cell. The receptors function as ligand-gated channels. Acetylcholine is broken down by the enzyme acetylcholinesterase, which is also localized on the external surface of the post-synaptic cell.

repeated over and over as an action potential is propagated along the full extent of the axon.

The electrical signal is relayed between cells in the nervous system at specialized regions of contact called synapses. Synapses occur between two neurons as well as between neurons and effector cells such as muscle cells. The latter type of synapse is referred to as a neuromuscular junction. Regardless of type, the mechanism of information transfer across a synapse is basically the same. The signal is relayed by means of a chemical neurotransmitter, molecules of which are packaged in membrane-bound vesicles in the pre-synaptic terminal (*Figure 1*). When an action potential reaches the pre-synaptic terminal, the depolarization activates calcium channels in the pre-synaptic membrane and the subsequent influx of calcium ions leads to the exocytotic release of neurotransmitter. The transmitter diffuses to the post-synaptic cell and interacts with specific receptors on the surface of that cell. Receptors function as ligand-gated ion channels that are activated in response to binding of the specific neurotransmitter molecules. Opening of the receptor channels initiates ion currents and results in the local depolarization of the post-synaptic cell to produce a synaptic potential. Generally, the size and duration of a synaptic potential reflects the amount of transmitter released by the pre-synaptic terminal. By depolarizing the post-synaptic cell above threshold, the synaptic potential can trigger the generation of an action potential to continue the signaling one step further in the neural pathway.

1.3 Electrophysiological methods in *Drosophila*

Several different preparations have been used in electrophysiological studies of excitable tissues in *Drosophila* embryos, larvae, and adults. Each system provides different types of information and has different advantages or disadvantages. Together, the various preparations enable the phenotypic effects of mutants to be characterized in various cell types and at different stages of development.

Neurons in the thoracic ganglion of adults were first recorded intracellularly by Ikeda (1,2). Action potentials have also been recorded in the cervical giant fibers, a pair of axons that run between the brain and the thoracic ganglion (3–5). Intracellular recordings from neurons *in situ* is technically difficult and has not been used extensively to characterize mutants. Extracellular recording provides less information about the amplitude and waveform of action potentials but can be used in some cases to monitor electrical activity in nerves. The larval nerves are particularly well-suited for this type of recording and the method has been used to characterize a number of different mutants (6–10).

The large size of the dorsal longitudinal flight muscles (DLMs) of adults and the medioventral longitudinal muscles of larvae faciliates intracellular recording from these cells. Defects in nerve activity, synaptic transmission and muscle response can be analyzed by recording the muscle response to nerve stimulation (5–9,11–16). In adults, the giant fiber pathway (GFP) innervates the DLMs (5). This pathway comprises several steps including sensory inputs to the giant fiber in the brain, the giant fiber itself, an interneuron, the DLM motorneuron and the DLMs. Muscle action potentials can be evoked by direct injection of depolarizing current via an intracellular electrode to distinguish defects in the muscle membrane itself from upstream defects elsewhere in the motor pathway (e.g. 11,14). As in other invertebrate species, the muscle action potentials in *Drosophila* differ from neuronal action potentials in that the inward current is carried by calcium ions rather than sodium ions.

In larvae, the motor axons within nerve bundles are readily accessible and can be stimulated directly to trigger a muscle response (15). Under physiological conditions, the synaptic potential in larval muscles does not evoke all-or-nothing action potentials. This is advantageous because the process of synaptic transmission that leads to the synaptic potential (referred to as an excitatory junctional potential) can be monitored without the masking effects of muscle action potentials (15,16).

Voltage-clamp experiments are performed to measure the amplitude of particular ion currents and to characterize the kinetics of activation and inactivation of these currents in response to abrupt shifts in membrane potential. The experimental technique usually requires a cell to be impaled with two electrodes and is most readily applied to relatively large cells. In *Drosophila*, only the currents in the pupal and adult DLMs (14,17,18) and the larval muscles (19,20) have been studied by this method.

An even more refined technique is now available to record currents flowing through single ion channels under voltage-clamp conditions. This recently developed patch-clamp technique employs extracellular pipette electrodes that can form an extremely tight seal with the plasma membrane to allow measurements of the currents that flow even through a single ion channel in the membrane patch under the pipette tip (21). Furthermore, with the cell still attached to the electrode, the membrane patch can be ruptured to gain access to the inside of the cell permitting intracellular recordings. In this 'whole cell' clamp configuration the total currents flowing through the entire cell membrane can be measured. This recording method does not require large cells and thus alleviates the technical difficulties associated with other types of electrophysiological measurements from small neurons in *Drosophila*. However, the patch-clamp technique works most effectively for cultured cells *in vitro* that are relatively free of connective tissue and able to form a tight seal with the patch electrode. Thus, the technique has so far been applied in *Drosophila* to primary cultures of dissociated neurons from the central nervous system (CNS) of larvae (22,23) and to embryonic neurons and myotubes derived from cultures of gastrulating embryos (24–26).

Electrophysiological studies of the compound eye have relied primarily on the electroretinogram (ERG), which is basically an extracellular

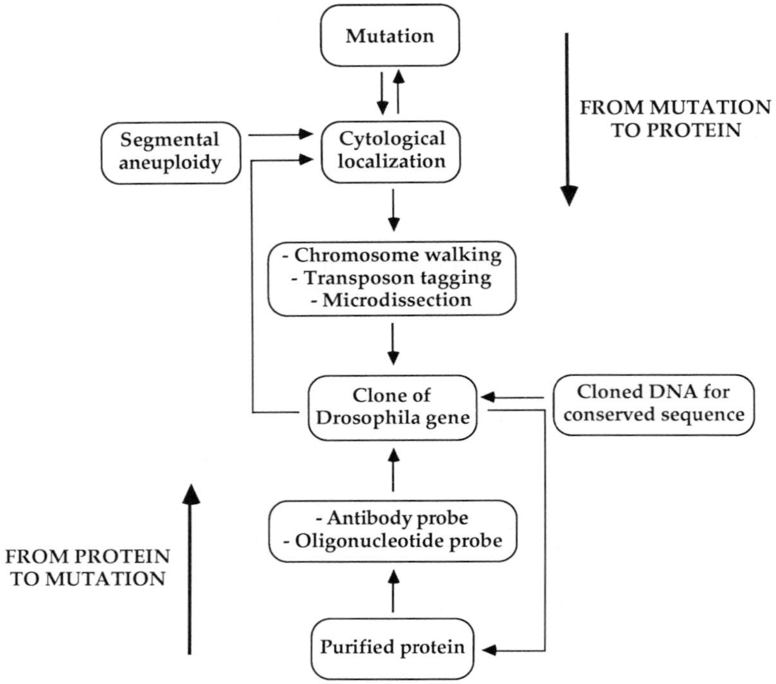

Figure 2. A flowchart diagramming the inter-relationships among the various experimental methodologies available in *Drosophila* for the genetic and molecular analysis of proteins that affect neural signalling. See text for details.

recording method. The ERG records the potential changes evoked by a light stimulus between an electrode placed on or just beneath the cornea of the eye and a reference electrode placed elsewhere inside the fly (27,28). One component of the ERG, a sustained corneal negative wave, is a summed response that corresponds primarily to depolarization of the photoreceptor cells. ERG recordings can be obtained with relative ease and the method is useful for rapid screening of mutants. However, to measure receptor potentials and to study the electrical properties of photoreceptor membranes it is necessary to carry out more difficult intracellular recordings from individual photoreceptor cells (29,30).

1.4 Strategies for cloning genes in *Drosophila*

One of the features that makes *Drosophila* as attractive for molecular neurobiology as it is for other areas of modern biology is the variety of strategies available to clone genes of interest (cf. ref. 31). Although some of these cloning methods can be applied equally well in other organisms, several methods exploit the kind of detailed genetic and cytological analyses that can best be performed in *Drosophila*. Below we present a brief overview of the different strategies used to clone the genes discussed in this review (*Figure 2*).

1.4.1 From protein to gene

In many instances, information is first acquired about a protein of interest before the gene encoding it is cloned. For example, once a protein is purified, a partial amino acid sequence can be obtained and an oligonucleotide fragment specifying that amino acid sequence can be synthesized. The oligonucleotide can then be used as a probe to screen genomic or cDNA libraries to isolate the desired gene. This approach has been used extensively for cloning vertebrate genes encoding a variety of ion channel species (see Chapter 2). If the cloned gene contains sequences that are conserved in evolution, the clones obtained first in other organisms can be used as probes to isolate the corresponding genes from *Drosophila*. Genes encoding rhodopsin (32,33; see Section 3.1), sodium channel (34; see Section 4.1) and an acetylcholine receptor sububnit (35,36; see Section 2.3) have recently been cloned from *Drosophila* in this manner. An alternative procedure for cloning a gene encoding a particular protein is to generate antibodies against the purified protein and use these antibodies to screen expression vector libraries. This procedure was used to obtain cDNA clones for choline acetyltransferse in *Drosophila* (37; see Section 2.2).

The generation of monoclonal antibodies against neural antigens in *Drosophila* enables detection of proteins whose identity is unknown but whose spatial or temporal distribution suggests functions of interest. The antibodies can be used to clone the relevant genes either by screening expression libraries or by using them to first purify the proteins so that a partial amino acid sequence can be derived. The latter approach has been used to clone *Drosophila* genes that are involved in neural development (e.g. 38,39).

1.4.2 From gene to protein

Perhaps the most useful feature of *Drosophila* for molecular studies is that a gene identified only by its mutant phenotype can be cloned by any of several methods. This circumstance permits the isolation and analysis of genes encoding proteins that are of functional interest but that have not previously been identified or purified in any organism. These cloning strategies all take advantage of the fact that genes in *Drosophila* can be mapped to a precise physical location because of the superb cytological resolution of their banded polytene chromosomes. A large number of stocks with various chromosomal rearrangements including deletions, duplications, and translocations that enable localization of a mutant gene to within a few chromosome bands already exists (40) and it is relatively easy to generate additional rearrangements as they are needed.

After a gene has been cloned, the amino acid sequence of the encoded product can be determined and compared with the sequences of other proteins in computer databanks to infer possible functions. The cloned genes can also be used in bacterial expression systems to obtain large amounts of purified gene product for the generation of monoclonal and polyclonal antibodies (41).

One cloning method takes advantage of cloned DNA segments previously isolated in *Drosophila* that happen to map to chromosomal sites in the immediate vicinity of the desired gene. These DNA probes can then be used to initiate a *chromosome walk* to the gene by the reiterative isolation of overlapping clones from a genomic library (42). This strategy was utilized to clone the *Sh* locus (43–45; see Section 5.1.1) and the same approach is being used to clone the *nap* locus (see Section 4.2.1). In general, this technique is useful when the desired gene is located to within about 200 kb of an available DNA probe.

If a gene of interest is not located close enough to a previously cloned site to make chromosome walking feasible, *chromosome jumping*, by use of chromosome rearrangements such as deletions, inversions or translocations, can sometimes overcome the problem by moving the gene closer to a chromosome region that has already been cloned (42). The existence of an inversion broken at one end in a previously cloned gene and broken at the other end in *eag* made the cloning of this gene possible (46; see Section 5.1.2).

Another approach, limited to some extent by the technical difficulty involved but enabling clones from virtually any small chromosome segment to be isolated, is *microdissection* of the appropriate segment from polytene chromosomes. Excised DNA is ligated into a phage vector to produce a mini-library of cloned DNA from the dissected region (31). The mini-library can then be screened for a particular gene or phage which can be used as a probe to initiate a chromosome walk in a standard genomic library.

The technique of *transposon tagging* has become one of the most generally useful and widely applied methods of cloning genes in *Drosophila*. The method depends upon the occurrence of mutations caused by the insertion of a transposable element in or near the gene of interest. The transposable element then provides a molecular handle to retrieve the DNA sequences that flank the insertion site from a genomic library. The most useful transposable element for this purpose in *Drosophila* is the P element (see ref. 47 for a review). Crosses between appropriate strains cause the otherwise stable P elements to become mobilized with a very high freqency in the germ line of the hybrid offspring, a condition known as *hybrid dysgenesis*. Germ line mobilization of the P element in dysgenic individuals leads to insertion of the element at new chromosomal sites and hence the production of many new mutations among the progeny of dysgenic flies. The mutation rates vary widely from gene to gene and some genes appear to be completely refractory ($0/10^6$) to P-element insertion (47). Therefore, the primary limitation in using P-element mutagenesis as a means of transposon tagging is the intrinsic susceptibility of a given gene to P-element insertion. However, as demonstrated by the cloning of *para* (48; see Section 4.2.2), once an appropriate insertional mutation has been found it is relatively straightforward to clone the gene.

1.5 Isolation of mutants

A number of strategies are also available to produce mutations in *Drosophila* that alter various aspects of nervous system function. In some cases, the goal has been to isolate mutations in specific genes that are known to encode proteins of interest. The isolation of mutations in genes encoding choline acetyltransferase (*Cha*) and acetylcholinesterase (*Ace*) are examples (49,50; see Sections 2.1 and 2.2). The genes that specify these enzymes were localized cytologically before mutants or clones were available for either gene. Lethal mutations, including those affecting *Cha* and *Ace*, were then generated in the appropriate chromosome locations. This cytological mapping was accomplished by means of a technique called segmental aneuploidy (51). The method is based on the observation that for many enzymes in *Drosophila* the level of activity is proportional to the number of copies of the gene encoding that enzyme that are present. Thus, flies carrying a single copy of the gene will have only 50% as much activity for the enzyme as do normal flies carrying the usual two doses. A fly with three copies of the gene will have 150% the normal activity. The particular chromosome region that includes the structural gene encoding an enzyme of interest can thus be localized even when no mutations of the gene exist. A special set of translocation stocks is available that enables investigators systematically to create such segmental aneuploids, that is, flies carrying three copies or one copy of any small chromosome region (51). By performing enzymatic assays on the segmental aneuploids, the chromosome segments containing the structural genes for choline acetyltransferase and acetylcholine esterase were identified and recessive lethal mutations mapping to the appropriate regions were generated (49,50).

A different type of genetic approach can be used when the identity of the proteins essential to the process under investigation and the location of the relevant genes are unknown. Mutations can be induced by chemical mutagens such as ethylmethane sulfonate (EMS), irradiation, or mobilization of transposable elements. Rather than screening for mutations affecting known proteins, the goal in this case is to recover mutations that disrupt developmental, physiological and behavioral processes as a means of identifying the genes and eventually the proteins that govern a given process. For example, to identify mutations affecting the structure and function of ion channels in *Drosophila*, mutations affecting membrane excitability were sought by screening collections of various behavioral mutants for those with electrophysiological defects (52). As described above, once a mutation has been isolated and mapped it is possible to clone the gene to identify the protein altered by the mutation.

Generally, once the first mutant allele of a gene has been found it is fairly straightforward to identify additional mutant alleles of the same gene because they fail to complement (i.e. restore the wild-type phenotype) the existing mutant allele in a heterozygous fly. The analysis of multiple mutant alleles of a gene is often critical in obtaining a complete assessment of the mutant phenotypes associated with that gene. In addition, mutant alleles associated with chromosome rearrangements or insertion of a transposable element that physically disrupts the gene can be instrumental in cloning the gene and in localizing its extent on a molecular map.

1.6 Germ line transformation and *in-vitro* mutagenesis

Another extremely valuable attribute of *Drosophila* is that techniques have recently been developed that enable investigators to introduce cloned genes back into the genome of flies (53). These methods again rely on the use of transposable P elements. A gene of interest is cloned into a P-element vector, which is then injected into a pre-blastoderm embryo. Some of the germ line cells of the injected embryos become 'transformed' with the injected DNA; the P-element vector and the inserted gene(s) it carries are stably inserted in the germ line and chromosomes carrying the newly inserted DNA can be transmitted to subsequent generations. Even though the genes become inserted at chromosome sites that differ from their normal locations, they are capable of functioning properly in their new location (e.g. 54–56). By *in-vitro* mutagenesis a cloned gene can be modified at any site or in any way the investigator chooses before it is reinserted in the genome (e.g. 57). These techniques are extremely powerful in elucidating the mechanisms that regulate the expression of particular genes and in analyzing structure–function relationships of a given gene product.

2. Molecular and genetic analyis of the cholinergic system

In *Drosophila*, as in other insects, acetylcholine (ACh) appears to be a major excitatory neurotransmitter in the CNS (58–61) (*Figure 2*). The biosynthetic enzyme, choline acetyltransferase (ChAT), and the degradative enzyme, acetylcholinesterase (AChE), as well as ACh are found at high levels in the *Drosophila* CNS throughout most of the life cycle (62,63). Acetylcholine receptors (AChRs) have also been detected in the *Drosophila* CNS by binding of α-bungarotoxin (BuTx) (64–66) and by ACh-mediated opening of ion channels in cultured neurons (22,67). Mutations in the genes encoding ChAT and AChE have been isolated

(49,50,68) and these genes have also been cloned (69,70). In addition, genes that appear to specify different subunits of the *Drosophila* AChR have recently been isolated (35,36,71,72). A number of experimental tools are thus available in *Drosophila* to manipulate the components of cholinergic synapses to analyze the role of these components in development, behavior and physiology.

2.1   *Cha*: the gene for choline acetyltransferase

The gene encoding ChAT was cytologically localized to region 91B-D on the polytene chromosomes by the technique of segmental aneuploidy (49). A group of 23 recessive lethal mutations mapping to this chromosome segment was isolated following chemical mutagenesis. In the heterozygous condition, four of these mutants had only half the normal level of ChAT activity (49). All four of these mutants behaved as alleles of a single gene, which has been named *Cha*. Two of the *Cha* alleles are unconditionally lethal, whereas the remaining two are temperature-sensitive (ts): they are viable if raised at 18°C but lethal when grown at 30°C. The lethal period for both classes of alleles is late in embryogenesis. Embryos homozygous for a Cha^{ts} allele have no detectable embryo activity when raised at 30°C. When grown at 18°C, Cha^{ts} mutants survive to adulthood but have much less than the normal ChAT activity. If these flies are shifted to 30°C as adults, the remaining ChAT activity is gradually eliminated over a period of several days. Loss of enzyme activity is paralleled by the appearance of an abnormality in a synaptic component of the ERG, defects in male courtship behavior and, with longer exposures to the high temperature, paralysis and death (49).

Prior to the onset of paralysis, exposure of Cha^{ts} homozygotes to 30°C results in electrophysiological abnormalities in the GFP, which enabled the demonstration that one of the synapses in this pathway is most likely cholinergic (73). Previous studies indicated that the GFP mediates the escape response of adults and drives the flight muscles through a trisynaptic pathway; the first synapse is electrical while the second and third are chemical (5). The transmitter at the third synapse, which is the neuromuscular junction, is most likely L-glutamate (16). Transmission at the second synapse, between the peripherally synapsing interneuron (PSI) and the individual DLM motorneurons (DLMn) is altered in Cha^{ts} mutants. Therefore, the transmitter at this synapse is most likely ACh. Exposure of Cha^{ts} mutants to 30°C for 1–1.5 days, which was insufficient to produce paralysis, caused the loss of the DLM response to GFP stimulation (73). Experiments to determine the site of failure in the pathway specifically pointed to the synapse between the PSI and the individual DLMns. Interestingly, when the DLM response to activation of the GFP was blocked in Cha^{ts} mutants, a separate neural circuit that drives the DLMns for flight remained intact. Thus, the integrity of the flight circuit, which involves a putative central command generator connected via a different pathway to the DLMns, apparently does not require cholinergic function (73). The GFP shows functional deficits in Cha^{ts} flies even when they are raised and maintained at permissive temperatures (73). Even at ChAT levels that are about 80% of normal, disruptions of the GFP become evident when the circuit is stressed by repetitive stimulation. The physiological experiments with Cha^{ts} demonstrate the utility of genetic methodology in identifying the transmitter at a particular synapse and in experimentally manipulating the available levels of transmitter.

ChAT was purified from *Drosophila* heads and a collection of monoclonal antibodies directed against this protein was generated (74,75). A mixture of three of the monoclonal antibodies was used to probe a head-specific cDNA expression library resulting in the isolation of 14 positive clones (37). *In-situ* hybridization to polytene chromosomes revealed that only one of these cDNA segments derives from 91B-D, the chromosome region that corresponds to the cytological location of *Cha*.

This cDNA clone is about 2.5 kb in length and has a coding region spanning 2190 nucleotides followed by a 3'-non-coding region 284 nucleotides in length. The clone appears to be incomplete at the 3' end since it lacks a poly[A]$^+$ tail. The coding region has a capacity to specify 728 amino acids, which is unexpected since this is 50–100 amino acids more than is required to account for the molecular weight (67 000) of the active enzyme. In addition, the 5' region of the cDNA sequence does not contain a methionine residue for the initiation of translation, indicating that the cDNA is incomplete at the 5' end as well (37). These results suggest that the polypeptide specified by the *Cha* mRNA is a large inactive precursor from which the active ChAT is derived. Furthermore, the *Cha* mRNA detected on Northern blots is 2.2 kb larger than the cDNA clone that was isolated, leaving the location and role of these additional bases in the message still unaccounted for (37). Since the cDNA clone was incomplete at the 3' end, some of these bases may be located in the non-coding region at this end. The cDNA clone could also be incomplete by as many as 2.0 kb at the 5' end. If so, it would be unusual to have a 5' sequence of this length that was not translated into protein (however, see the description of the *Ace* mRNA below). If, instead, the extra 2.0 kb of the *Cha* mRNA represents additional coding information at the 5' end of the message, it would be further evidence that the initial protein product of the *Cha* locus is a large inactive precursor to the active enzyme. The results of the molecular analysis of the *Cha* locus thus support the previous suggestion based on biochemical studies that production of active ChAT in *Drosophila* requires post-translational modification by limited proteolysis of a larger precursor (63,75).

2.2 *Ace*: the gene for acetylcholine esterase
AChE is responsible for the hydrolysis of ACh at the surface of the

post-synaptic cell terminating the interaction of this transmitter with the AChR. Mutants that abolish AChE activity enabled the role of this enzyme in the normal development and function of the *Drosophila* nervous system to be studied and led to the molecular isolation of the structural gene. The gene encoding AChE was localized by the analysis of segmental aneupoloids to the cytological region 87E1-5 (50). Subsequently, a group of allelic recessive lethal mutations in this gene (*Ace*) were identified. Homozygotes died as embryos and had no detectable AChE activity (50). Additional alleles were recovered that were temperature-sensitive for AChE activity (68). Mosaic analysis of lethal *Ace* alleles indicated that although the enzyme is present throughout the *Drosophila* CNS (comprising optic lobes, brain and thoracic ganglion) AChE appears to be required for survival principally in the posterior midbrain (68). Very few surviving adult mosaics were found that had even small unilateral clones of mutant tissue in this region and none were found that had bilateral clones of mutant tissue in the region. In contrast, unilateral and even bilateral clones of mutant tissue were found at reasonable frequencies in other portions of the CNS. The viable mosaics did, however, manifest defects in visual physiology, optomoter response, and courtship behavior depending on location of the mutant tissue. In addition, the neuropil in the mutant patches displayed abnormal morphology including reduced volume, compacted appearance, apparent disorganization of axons and, in some cases, degeneration (68). These results demonstrated that AChE activity is essential in *Drosophila* for the normal development and maintenance of some neural structures.

The cytological location of *Ace* is close to that of the well-characterized *rosy* locus at 87D11-13. The *rosy* locus and the surrounding regions were cloned in a chromosome walk that encompassed over 300 kb (42). The breakpoints of chromosome deletions that defined the location of *Ace* were mapped at the DNA level by Southern hybridization analysis and a region of 40 kb that included the *Ace* locus was delimited (76). A fragment of genomic DNA from within this region was used to screen embryonic and pupal cDNA libraries resulting in the isolation of a set of overlapping clones that appeared to be derived from the same gene (70). The combined length of the cDNA clones corresponds to a transcript of about 4.2 kb. DNA sequence analysis revealed a long open reading frame of 1950 bases. The predicted amino acid sequence confirms that the encoded protein is AChE. In the correct translation frame the consensus sequence Phe-Gly-Glu-Ser-Ala-Gly, of the active site of vertebrate cholinesterase is found (70,77). The amino acid sequence of *Drosophila* AChE is extensively homologous to the AChE encoded by a cDNA clone isolated from *Torpedo*; 31% of the amino acids from *Drosophila* are identical to *Torpedo* and the alignment between the two sequences requires few insertions or deletions (70).

An unusual feature of the cDNA clones for *Drosophila* AChE is a 5' non-coding leader that is at least 1.0 kb long (rather than the 40–80 bases typical for *Drosophila*) and contains six ATG codons preceding the presumed initiating ATG codon. The upstream ATG codons are, in principle, capable of initiating translation in five open reading frames of between 1 and 43 amino acids. The unusual structure of the 5' leader suggests that it may play a role in regulating the expression of AChE at the translational level (70). In this context, it is of interest to recall that the *Cha* mRNA may also possess an unusual 5' leader sequence. Perhaps the genes encoding both the synthetic and degradative enzymes for ACh share similar translational control mechanisms.

In other insects several distinct forms of AChE occur. Only a single gene encoding AChE has been detected in *Drosophila* by genetic and molecular criteria (50,70). Differential processing of transcripts from this one locus is one mechanism by which different subunit types could be produced. On Northern blots of pupal poly[A]$^+$ RNA probed with one of the AChE cDNA clones, two prominent bands are detected at 4.5 and 4.8 kb with fainter signals above and below these bands. The same transcripts are detected at other developmental stages although they are less abundant. The presence of these multiple RNA species suggests the possibility that alternative processing of the AChE transcript may be occurring although this possibility still awaits confirmation.

2.3 A gene for an acetylcholine receptor subunit

The *Drosophila* CNS is a rich source of a BuTx-binding component with properties similar to those of the vertebrate nicotinic AChR (64–66). Furthermore, single-channel currents from ACh-gated channels can be readily recorded in primary cultures of dissociated neurons from the larval CNS (22,67). Thus, it was expected that *Drosophila* would have an AChR with structural similarities to the vertebrate receptor.

Clones encoding a putative subunit of the *Drosophila* AChR were isolated using a cDNA probe from the γ subunit of *Torpedo* electroplaque (35,36,71). A cDNA library prepared from poly[A]$^+$ RNA of heads (35, 71) and a genomic library (36) were screened under conditions of low stringency hybridization. The cDNA and the genomic DNA segments isolated from these libraries apparently represent the same AChR subunit-related gene (35,36). The cytological location of this gene by chromosomal *in-situ* hybridization is on the third chromosome at 64B. No mutations affecting the gene encoding this putative AChR subunit have yet been isolated, although it should be feasible to do so in screens analogous to those used to find *Cha* and *Ace* mutations.

From nucleotide sequence analysis of the AChR-related clones, the mature *Drosophila* protein is predicted to be 497 amino acids in length and to have several features in common with vertebrate nicotinic AChR subunits (35,36). These features include the distribution of hydrophobic regions, an amphipathic α-helix and two cysteine residues near the N terminus, which are thought to be essential in formation of the tertiary structure of AChR subunits. The structural features of the polypeptide make it seem very likely that it does represent a subunit of the *Drosophila* AChR, although this conclusion has not yet been demonstrated by any functional assays.

Comparison of the amino acid sequence of the *Drosophila* polypeptide with other AChR subunits from *Torpedo* electroplaques reveals that it shares the highest homology (41–44%) with the α and β subunits (35,36). The homology was least with the γ subunit (33–34%), even though DNA probes for this subunit were used to isolate the *Drosophila* clones. Among the various comparisons made, the deduced *Drosophila* polypeptide shared the highest homology (46%) with a putative rat neuronal α subunit, which was cloned from PC12 cells (78). Despite its high homology with vertebrate AChR α subunits, the *Drosophila* polypeptide lacks the two consecutive cysteines at amino acids 194 and 195, which are a characteristic feature of the ligand-binding site of all other α subunits studied. This raises the possibility that there are still other AChR subunit-encoding genes that remain to be identified in *Drosophila*. In fact, isolation of DNA segments that apparently encode another *Drosophila* AChR subunit with placement of cysteine residues that correspond to the vertebrate α subunit was recently reported (cited in ref. 72).

Expression of the *Drosophila* AChR-related gene at 64B was studied by Northern blot and tissue analysis *in situ*. A single 3.2 kb transcript was detected at low levels in 4 h embryos and at much higher levels again during late pupation (35,36). These times of expression coincide with periods of extensive neuronal differentiation during *Drosophila* development. Consistent with the belief that the transcript specified by this AChR-related gene encodes a functional subunit of the *Drosophila* AChR, the tissue *in-situ* studies revealed that the transcript localizes exclusively to the CNS at all stages of development (36).

Further studies of the putative *Drosophila* AChR genes, especially once mutations have been isolated, should provide additional information about the structure, function and evolution of the AChR.

3. Molecular and genetic analysis of phototransduction

The process of phototransduction involves the conversion of the energy of absorbed photons of light into a receptor potential in photoreceptor cells (79). Since absorption of even a single photon can evoke a discrete receptor potential (e.g. see ref. 30), an amplification process must be involved. The first step in the biochemical cascade that leads to the receptor potential is the light-induced conformational change in the visual pigment, rhodopsin. It consists of an apoprotein, opsin, covalently bound

24 Molecular neurobiology

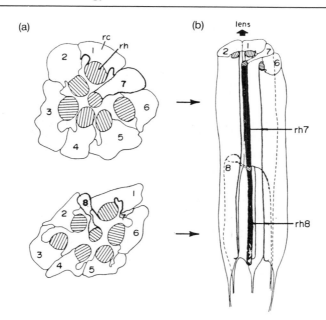

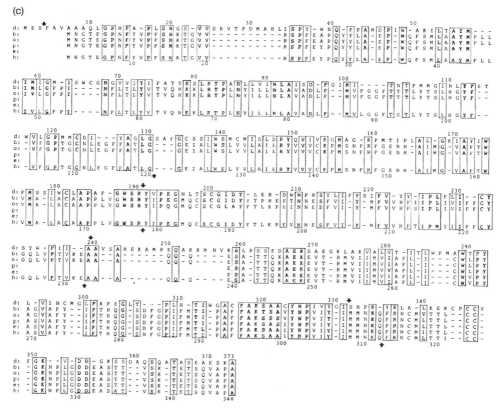

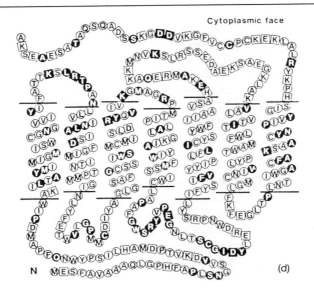

Figure 3. Schematic diagram showing the positions of the eight retinula (photoreceptor) cells and their rhabdomeres within an ommatidium. (**a**) Cross-sectional views at the levels indicated by the arrows. In (**b**) retinula cells R3–5 are omitted for clarity; rc, retinula cell; rh, rhabdomere. Reproduced with permission from ref. 138. (**c**) Comparison of *Drosophila* opsin 1 amino acid sequence, encoded by the *ninaE* locus, with the mammalian opsin family. b, bovine; o, ovine; p, porcine; e, equine; h, human. Reproduced with permission from ref. 32. (**d**) A proposed structure for *Drosophila* rhodopsin. Black solid circles indicate identities between the *Drosophila* and bovine sequences, shaded circles denote conservative amino acid changes and open circles represent positions where the *Drosophila* and bovine sequences differ. Reproduced with permission from ref. 33.

to a chromophore, 11-*cis*-retinal, which is isomerized to 11-*trans*-retinal by absorption of a photon. This conformational change triggers a set of reactions that results ultimately in the activation (or closing in the case of vertebrates) of a specific set of channels in the photoreceptor membrane. The intermediate steps between absorption of light and channel activation that lead to the receptor potential involve a G-protein-dependent second messenger system but all the molecular details of this process have not yet been worked out (79).

Each of the 800 ommatidia in the compound eye of *Drosophila* contains three distinct classes of photoreceptor cells, distinguishable by their morphological arrangement, absorption spectra, and function (80). There are six peripheral photoreceptor cells (R1–R6) in each ommatidium and two central photoreceptor cells (R7 and R8) (*Figure 3*). Cells R1–R6 contain the major form of rhodopsin, which has maximal absorption at 480 nm. They form the high sensitivity system and operate at low ambient light intensity. R7 and R8 each appear to contain minor forms of rhodopsin; a UV-sensitive and a blue non-adapting pigment, respectively. They form the high acuity system and are capable of detecting refined patterns and polarized light. The genes encoding the different forms of opsin have now been cloned, enabling molecular comparison of their gene products.

3.1 The R1 – R6 opsin gene: *ninaE*

The major form of opsin is the product of the *ninaE* (neither inactivation nor afterpotential E) locus (32,33). Mutations in this gene were originally identified among a collection of mutants defective in the phototransduction process that displayed characteristic ERG abnormalities (29). The amount of visual pigment in photoreceptors R1 – R6 is drastically reduced in *ninaE* mutants but the amount of pigment in cells R7 and R8 is unaffected (81). Moreover, changes in dosage of the wild-type allele in segmental aneuploids result in correlated changes in the content of visual pigment in cells R1 – R6. From these results it was proposed that *ninaE* encodes the opsin found in photoreceptors R1 – R6 (81).

Two groups independently isolated *Drosophila* genomic clones (32,33) based on weak homology to available bovine opsin cDNA probes (82). In both cases, the *Drosophila* clones were localized by chromosomal *in situ* hybridization to 92B4-11, the region that contains *ninaE*. That the cloned DNA segments correspond to the *ninaE* gene itself was indicated by the fact that two *ninaE* mutations, a small deletion and an insertion, each had alterations detectable by Southern blot analysis within the cloned region (32). Northern blots revealed that in normal flies *ninaE* expresses a single, abundant 1.7 kb transcript. The level of this transcript is reduced by more than 99% in the two *ninaE* mutants mentioned above and in other mutants that lack rhabdomeres, the rhodopsin-containing structures of photoreceptors (32,33).

DNA sequence analysis shows that *ninaE* encodes a polypeptide chain (designated Rh1) of 343 amino acids. Direct comparison of the *Drosophila* and bovine opsin nucleotide sequences reveals scattered regions of homology (32,33). The detectable hybridization between the *Drosophila* and bovine sequences is attributable to two small homologous regions with the best match being a string of 44 bases with 10 mismatches. The similarity between *Drosophila* and mammalian opsins is greater at the amino acid level (*Figure 3*). When the *Drosophila* amino acid sequence is aligned for maximum homology with mammalian opsin sequences (bovine, ovine, equine, porcine, and human), 36% of the amino acids in *Drosophila* are identical with any given member of the mammalian family (32). If substitutions of amino acids with similar properties are counted,

49% of the amino acids are conserved. Three clusters of strong homology are apparent. One of these regions corresponds to the retinal chromophore binding site; the other two sites of strong homology may also identify regions of important structural or functional significance. The overall structural configuration of *Drosophila* and bovine opsins appears to be quite similar; each polypeptide contains seven hydrophobic, putative transmembrane segments (32,33).

The cloned Rh1 gene has been reintroduced into the *Drosophila* germ line by P-element mediated gene transfer (C.S.Zuker, personal communication). The introduced gene complements the *ninaE* mutation, showing that Rh1 is expressed and functions normally when reintroduced into the *Drosophila* genome. It is now possible to extend these studies by using *in vitro* site-directed mutagenesis to mutate selected amino acids and regions in the rhodopsin molecule and then analyzing the properties of the altered protein following reintroduction of the mutant genes into flies. Such studies should help elucidate the functional significance of the different structural features of the rhodopsin molecule as well as the interactions between the photoactivated rhodopsin and other members of the phototransduction cascade such as transducin and rhodopsin kinase.

3.2 Genes for minor forms of opsin

3.2.1 The R8 opsin gene

The different absorption spectra of the photopigments found in the three classes of photoreceptors suggest that they are encoded for by different opsin genes. To identify other putative opsin genes, probes from the *ninaE* gene were used to screen *Drosophila* libraries at reduced stringency for cross-homologous sequences. One clone, Rh2, isolated in this way defines an R8-specific opsin (83).

By *in-situ* hybridization Rh2 maps to chromosome position 91D1-2. None of the existing mutations affecting phototransduction is located in this region. However, DNA sequence analysis confirms that this clone encodes an opsin (83). The deduced polypeptide sequence contains all the structural features characteristic of opsin, including seven hydrophobic domains and the retinal-binding site in the seventh transmembrane domain. The 381-amino-acid Rh2 polypeptide shares 67% amino acid identities with the *ninaE* opsin. Using gene-specific probes, The Rh2 transcript was examined in normal flies and in mutants lacking rhabdomeres in photoreceptors R1–R6 (e.g. extreme alleles of *ninaE*) or R7 (*sev:* sevenless). The level of Rh2 transcript is not altered in these mutants suggesting that it is expressed primarily or entirely in R8. Tissue *in-situ* hybridization was used to show directly that Rh2 is transcribed specifically in R8 photoreceptors (83).

Detailed photochemical analysis *in situ* of the properties of the R8 rhodopsin has not previously been accomplished in *Drosophila*. Microspectrophotometric analysis has been precluded since the R7 rhabdomere is stacked on top of the R8 rhabdomere. Molecular genetic techniques are now being used to circumvent some of the difficulties in studying the minor photopigment specified by Rh2. The promoter that normally drives the transcription of Rh1 in photoreceptors R1–R6 has been linked *in vitro* to the coding sequence of Rh2 and the hybrid construct reintroduced by P-element mediated transformation into the genome of *ninaE* flies lacking the Rh1 opsin. Flies were thus produced that expressed the normal R8 photopigment in cells R1–R6. The photochemical properties of the R8 pigment and the behavior of the flies that are mis-expressing it are now being investigated (R.Feiler, W.Harris, K.Kirschfeld, C.Wehrhahn, and C.Zuker, personal communication).

3.2.2 The R7 opsin genes

Two additional R7-specific opsin genes were identified in similar fashion to Rh2. Low stringency screens of *Drosophila* genomic and cDNA libraries with *ninaE* and Rh2 probes failed to detect any other homologous sequences (84). However, more sensitive screens with oligonucleotide probes corresponding to two of the most highly conserved regions between *ninaE* and Rh2 opsins did result in the identification of a homologous gene, Rh3 (84). A DNA segment identical to Rh3 was also isolated by using rhodopsin oligonucleotide probes to screen a collection of genomic clones representing genes expressed specifically in the eye (85). By DNA sequence analysis, Rh3 was found to encode a 383 amino acid polypeptide with the distinguishing structural features of opsin. There are 130 and 125 amino acid identities, respectively, with *ninaE* and Rh2 opsins. No mutations in Rh3, which maps to cytological region 92D, have yet been found.

Rh3 was used in turn to screen an adult *Drosophila* head-specific cDNA library and a cDNA clone mapping to 73D3-5 was recovered (86). This cDNA segment defined a gene, Rh4, which specifies a 378-amino-acid polypeptide structurally similar to other opsins. Rh4 opsin is much more closely related to Rh3 opsin than to the *ninaE* or Rh2 opsin; it shares 72% amino acid identity with Rh3 opsin but only about 35% identity with other *Drosophila* opsins. Both Rh3 and Rh4 are transcribed specifically in R7 photoreceptors as indicated by tissue *in-situ* hybridization and Northern blot analysis of normal and *sev* flies (86). *In-situ* hybridization to adjacent sections through the eye with an Rh3 and an Rh4 probe, respectively, yielded the surprising result that the two R7 opsins are expressed in non-overlapping subsets of R7 cells; of 159 ommatidia examined all expressed either Rh3 (81/159) or Rh4 (78/159) and none were found that expressed both (86). It is of interest to note that two classes of photoreceptor cells with different spectral properties have been reported in other dipteran species.

The four opsin genes that have now been identified may be sufficient to account for all the spectral properties of the *Drosophila* compound eye. Further genetic and molecular analyses of these four opsin genes should help to elucidate the particular structural features that underlie their distinctive functions.

3.3 Genes affecting intermediate steps of phototransduction

Photoisomerization of visual pigments in both vertebrates and invertebrates initiates a series of intermediate steps that amplifies the original signal and leads to a change in the permeability of the photoreceptor membrane and the generation of electrical impulses. The molecular basis of these intermediate steps and the nature of the proteins involved can be studied in *Drosophila* by analyzing mutants that are defective in these steps. Examples of two such mutants, the molecular analyses of which are now under way, are transient receptor potential (*trp*) and no receptor potential (*norpA*).

3.3.1 The trp locus

Mutations of the *trp* locus are recessive and map on the third chromosome near the tip of the right arm. The first mutant allele appeared spontaneously in a highly inbred line (87). Additional mutant alleles were isolated following EMS mutagenesis (29). The mutant flies behave normally under low ambient light conditions but appear to be blind when the light level is high (29). The blindness is associated with an abormality in the process of phototransduction that is revealed in ERG or intracellular recordings (87,88). The receptor potential in *trp* mutants decays more rapidly than normal during illumination with bright light; the brighter the illumination, the faster the rate of decay (88). In addition, recovery in the dark from a previous light stimulus occurs abnormally slowly in *trp* mutants. The biochemical basis of the *trp* phenotype is not well understood but it does not arise from alterations in the amount or properties of rhodopsin in the photoreceptor cells (29). A detailed molecular analysis of the product encoded by the *trp* locus could therefore help elucidate its mode of action (89,90).

Because of the phenotype of *trp* mutants, it was expected that this gene would be transcribed into RNA only in the eye. Consequently, cloned DNA sequences that were expressed at least ten times more abundantly in heads than bodies were isolated by a differential screen of a *Drosophila* genomic library using poly[A]$^+$ RNA prepared from fly heads and bodies (89). Among 20 genomic DNA segments isolated in this way (89), one mapped by *in-situ* hybridization to 99C, the cytological region known to contain *trp* (90). Further evidence that this DNA segment included at least part of the *trp* locus was obtained from tissue *in-situ* hybridization experiments to determine the spatial distribution of transcripts specified by this cloned segment; the transcript was expressed predominantly in the eye and perhaps exclusively in the photoreceptor cells (89). Additional sites of expression of the transcript were in the simple eyes or ocelli, whose function is known also to be affected by *trp*.

The transcript detected by the putative *trp* clone on Northern blots is 4.2 kb in length and is first expressed in pupae within 24 h of eclosion and limited to adult heads thereafter (89). Final proof that this 4.2 kb RNA corresponds to the *trp* message was provided by the demonstration that a segment of genomic DNA that completely encompasses this transcript could rescue the *trp* mutant phenotype following P-element mediated germ line tranformation. It should now be possible to determine the molecular identity of the *trp* gene product and to begin to elucidate its function in phototransduction at a biochemical level.

3.3.2 The norpA locus

The *norpA* mutants were among the first group of ERG-defective mutants isolated (91,92). Over 40 mutant alleles of this locus have been found ranging in phenotype from a complete lack of the receptor response to stimulation by light, to only a moderate reduction in amplitude and a prolonged time-course of the receptor potential (29). When tested in a photomaze, flies mutant for a severe *norpA* allele are completely blind (29). There is no apparent effect of *norpA* mutations on the level of rhodopsin in photoreceptor cells or in the photochemical properties of this pigment (93,94). Instead, electrophysiological and spectrophotometric studies on temperature-sensitive *norpA* alleles led to the suggestion that *norpA* was specifically defective in an intermediate step of phototransduction subsequent to the light-activation of rhodopsin.

More recently, it has been shown that *norpA* alleles affect levels of phospholipase-C in the eye in a manner consistent with the idea that *norpA* is the structural locus for this enzyme (95). This result is of interest because phospholipid derivatives have been implicated as a second messenger in invertebrate phototransduction (96,97).

The *norpA* locus was mapped to the tip of the X chromosome in cytological region 4B6-4C1. Genomic DNA segments from this region encoding a head-specific transcript have now been isolated. This DNA segment corresponds in location to the site containing the insertion of a transposable element in a hybrid-dysgenesis-induced *norpA* allele. Analyses of cDNA clones that correspond to the head-specific transcript from this region are now under way (B.Bloomquist, R.Shortridge, C.Montell, H.Steller, G.Rubin, and W.L.Pak, personal communication). Along with the analysis of *trp*, further investigations of *norpA* may lead to a more complete understanding of the intermediate steps of phototransduction at a molecular level.

4. Molecular and genetic analysis of sodium channels

Sodium channels play the primary role in the generation and propagation of action potentials in neurons. Elucidation of their structure and function is thus central to understanding the molecular basis of neural signaling. Functional properties of sodium channels that still await detailed explanations in terms of the molecular structure of the protein include the selective permeability to sodium ions and the voltage-sensitive gating mechanisms for activation and inactivation of channel opening (98). In addition, the mechanisms that regulate the expression, post-translational modification, membrane distribution, etc., of sodium channels remain to be elucidated. Molecular analysis of sodium channels is now being carried out in several different organisms (see Chapter 3) but the use of genetics and germ line transformation in *Drosophila* offers important advantages because they permit the functional analysis of mutant channels *in situ*.

Two complementary approaches have been taken to initiate studies of the structure, function, and regulation of sodium channels in *Drosophila*. In one approach investigators are using cDNA or oligonucleotide probes to clone *Drosophila* genes that are homologous to the genes encoding the α subunit of vertebrate sodium channels. In the other approach, mutations affecting sodium channels are first isolated and characterized. The genes thus identified are being cloned to enable a molecular analysis of the proteins they encode. These two approaches are now beginning to converge as exemplified in the analysis of *para* (see Section 4.2.2).

4.1 A *Drosophila* homolog of the vertebrate sodium channel

A gene isolated by screening *Drosophila* genomic libraries at low stringency with a cDNA probe (34,99) or a synthetic oligonucleotide (Y.Hotta, personal communication) encoding a portion of the eel sodium channel has been shown to specify a polypeptide very similar to the vertebrate sodium channel. The amino acid sequence of the *Drosophila* protein was deduced from the nucleotide sequence of cloned genomic DNA and a partial cDNA clone. Like the α subunit of the vertebrate channel (*Figure 4*), the *Drosophila* protein contains four internally homologous domains, each about 300 amino acids long. Each of the homologous domains contains the six presumptive membrane-spanning regions (S1 – S6) identified in the eel and rat proteins (100,101). In general, there is a high degree of amino acid conservation between the *Drosophila* and vertebrate proteins within the four homologous domains (34). Regions S5 and S6, which are the most hydrophobic of the presumptive membrane-spanning regions, are the most conserved regions between the *Drosophila*

32 Molecular neurobiology

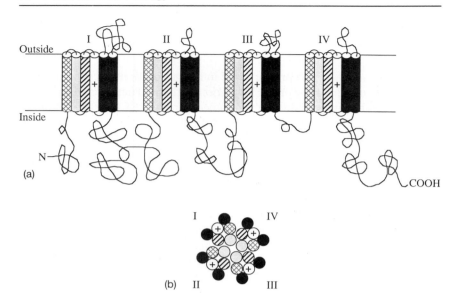

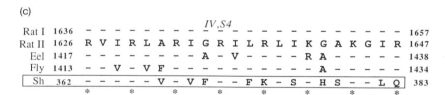

```
                            IV,S4
Rat I  1636  - - - - - - - - - - - - - - - - - - - - - -  1657
Rat II 1626  R V I R L A R I G R I L R L I K G A K G I R  1647
Eel    1417  - - - - - - - - A - V - - - - R A - - - - -  1438
Fly    1413  - - V - V F - - - - - - - - - - A - - - - -  1434
Sh      362  - - - - - V - V F - - F K - S - H S - - L Q   383
             *       *       *       *       *       *       *       *
```

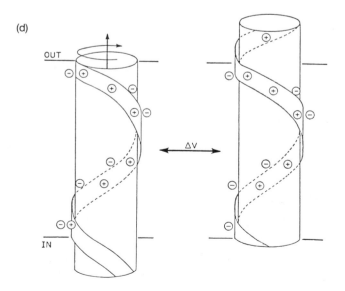

(a), (b), (c), (d)

Neurophysiology in *Drosophila* 33

and vertebrate proteins (34,100,101). The S4 regions in vertebrate sodium channels are characterized by a repeating sequence of a positively charged amino acid (usually arginine) alternating with two uncharged amino acids (*Figure 4*). These regions are postulated to serve as the voltage sensor (100) (*Figure 4*). All 24 positive charges in the S4 regions of the vertebrate protein are present at identical positions in the *Drosophila* protein (*Figure 4*), although the *Drosophila* protein contains two additional positive charges not found in vertebrates (34). The four homologous domains are connected to each other by three hydrophilic linker segments that are believed to be cytoplasmic (100,101). In contrast to the homologous repeats, the first two linker segments show very little amino acid conservation between *Drosophila* and vertebrate genes (34). The third linker segment, which has been proposed to be part of the inactivation gate (100,101), is highly conserved between the vertebrate and *Drosophila* genes (34).

The transcript from the *Drosophila* gene is larger than 10 kb. In very early embryos (0–9 h) two transcripts of different sizes are present suggesting the possibility of alternative processing. Only the larger of the two transcripts is evident in later embryos and pupae. Transcription of the gene in adults is negligible (34). It would be of interest to know whether

Figure 4. (a) A proposed membrane topology of the vertebrate sodium channel. The four homology units spanning the membrane are shown. Each of the six presumptive membrane spanning domains (S1–S6) in each homology unit is represented as a cylinder. (b) A cross-sectional view of a proposed arrangement of the transmembrane segments. The ionic channel is shown as a central pore surrounded by the four homology units. (a) and (b) are redrawn from ref. 101. (c) Alignment of the amino acid sequence of the S4 domain from the fourth homology unit of sodium channels from rat, eel, and *Drosophila*. Dashes represent amino acid identities with the rat II sequence. Positions of positively charged amino acids presumed to play a role in gating of the sodium channel are indicated with an asterisk. The amino acid sequence of the *Sh* protein (box), a probable component of potassium channels, also shares homology with this domain of the sodium channel polypeptide. The rat amino acid sequences are from ref. 101, the *Drosophila* sodium channel sequence is from ref. 34 and the *Sh* sequence is from ref. 128. (d) The sliding helix model suggesting how the S4 domains may be involved in voltage-dependent gating. The proposed transmembrane S4 helix is illustrated as a cylinder with a spiral ribbon of positive charge that moves in response to membrane depolarization. At the resting membrane potential (left), all positively charged residues are paired with fixed negative charges on other transmembrane segments of the channel and the transmembrane segment is held in that position by the negative internal membrane potential. Depolarization (right) reduces the force holding the positive charges in their inward position. The S4 helix is then proposed to undergo a screw-like motion through a rotation of approximately 60° and an outward displacement of approximately 5 Å. This movement leaves an unpaired negative charge on the outward surface to give a net charge transfer of 1. Reproduced with permission from ref. 139.

this gene does not need to be transcribed in adult neurons because a different gene directs the synthesis of sodium channels at this stage of development.

The sodium channel-homologous gene localizes to region 60D-E of the polytene chromosomes by *in-situ* hybridization (34). This location is distinct from that of the *sei* locus (60A-B), which was previously suggested to encode a component of sodium channels (102,103; see Section 4.2.4). None of the other mutations thought to affect sodium channels reside anywhere in this region. Thus, the role of the protein encoded by the cloned *Drosophila* gene is still not demonstrated at a functional level. To address this issue, sodium currents are being examined in cultured neurons from individual embryos homozygous for deletions that remove chromosome segments in the region of the gene (D.O'Dowd, S.Germeraad, and R.Aldrich, personal communication). So far, half of the relevant chromosome region has been examined but no discernible effects on sodium currents have been found. It is possible that the gene resides in that portion of the chromosome region not yet examined in deletion-bearing embryos or that a different gene is responsible for the production of functional sodium channels in embryonic neurons. It is known that other genes in *Drosophila*, including at least one identified by mutations, also encode polypeptides that share considerable amino acid similarity with the vertebrate channel (see Section 4.2.2).

4.2 Mutations affecting sodium channels

4.2.1 The nap^{ts} mutation

The *nap*^{ts} (no action potential, temperature-sensitive) mutation was recovered in a screen for temperature-sensitive paralytic mutants on the second chromosome (6). When homozygous *nap*^{ts} larvae or adults are shifted from 25 to 37°C, they become paralyzed within seconds. Normal mobility is regained almost immediately upon return to 25°C. Extracellular recordings of action potentials in *nap*^{ts} larvae revealed a failure in nerve conduction at 37°C (*Figure 5*). Upon lowering the temperature active conduction is restored. Excitatory junctional potentials recorded intracellularly from larval muscles in response to nerve stimulation are alike in *nap*^{ts} and normal larvae at 25°C but the response abruptly disappears in mutant larvae at elevated temperatures. Direct electrotonic depolarization of the nerve terminal could still evoke transmitter release and post-synaptic response in *nap*^{ts} larvae at 37°C, demonstrating a specific defect in the propagation of axonal action potentials (6).

Additional defects in nerve – membrane excitability are displayed by *nap*^{ts} even at nomally permissive temperatures: in comparison with wild-type larvae, the nerve refractory period is abnormally long and conduction is blocked at concentrations of tetrodotoxin (TTX) 4- to 5-fold lower (9); the behavioral phenotype and repetitive nerve activity associated

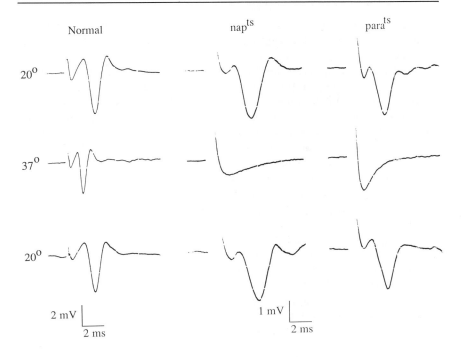

Figure 5. Temperature-dependence of action potential propagation in normal, nap^{ts}, and $para^{ts}$ nerves. Larval segmental nerves were cut near the ganglion and stimulated at that end. The compound action potential was recorded *en passant* by a suction electrode at some distance from the stimulation site. The exact waveform depends on the length of the nerve loop drawn up in the suction electrode. At 20°C, the compound action potential in the mutants is similar to that in normal flies. At 37°C, the compound action potential is still prominent in normal nerves. However, in nap^{ts} and $para^{ts}$, action potentials are completely absent at this temperature and only the stimulus artefact is recorded. The active response in the mutants recovers rapidly when the temperature is lowered again to 20°C. Reproduced from ref. 48.

with potassium channel mutants are suppressed in double mutants with nap^{ts} (8,104); dissociated neurons cultured *in vitro* are more resistant than normal to the cytotoxic effects of veratridine, a drug that causes persistent activation of sodium channels (105). In fact, nap^{ts} neurons are almost as resistant as normal neurons whose sodium channels have been completely blocked with TTX. These data are all consistent with the idea that nap^{ts} is altering sodium channels in some way.

Biochemical characterization of nap^{ts} by ligand-binding studies with [^{3}H]TTX or [^{3}H]STX (saxitoxin) lends further support to this interpretation (102,106). Binding properties of crude membrane extracts prepared from heads of nap^{ts} and normal flies were compared. Although the binding affinity for TTX and STX and the pH profile of binding activity are not altered in nap^{ts}, the number of specific binding sites is reduced by about 40%. These results were interpreted to mean that nap^{ts} is defective in regulating the number of sodium channels but does not alter their structure or function (102). However, other interpretations are not ruled out. For example, nap^{ts} could directly or indirectly alter the structure of sodium channels leading to impaired function and diminished stability in the membrane without altering the toxin binding site itself. Sodium channel variants with such altered properties have been isolated in a mammalian neuroblastoma cell line (107). To distinguish between the various possible interpretations of the *nap* phenotype it would be useful to examine toxin binding with different *nap* alleles and to perform binding assays with toxins that recognize sites on sodium channels different from the STX site.

Molecular analysis of the *nap* gene product will be essential for detailed understanding of its role in nerve excitation. Efforts to clone the *nap* locus are now in progress (M.Kernan and B.Ganetzky, unpublished data). The gene was first localized cytologically by generating a large number of radiation-induced deletions that uncovered the *nap* locus. These deletions delimited the locus to within several salivary bands in region 42A. The region thus defined is very close to a previously cloned actin gene. Using the actin clones as a starting point, a chromosome walk was initiated and approximately 120 kb of DNA have now been isolated from the *nap* region. By *in-situ* hybridization and Southern blot analysis the deletion breakpoint that defines the distal extent of the *nap* region has now been localized on the walk. In addition, the breakpoint of a translocation that fails to complement the mutant phenotype of nap^{ts} has also been located on the molecular map. This result suggests that the walk already includes DNA very close to if not within the *nap* locus itself. A molecular description of *nap* should therefore be forthcoming.

4.2.2 The para locus

The first temperature-sensitive paralytic mutation recovered identified the X-linked *para* (paralytic) locus (108). The first mutant allele, $para^{ts1}$, is recessive and causes the instantaneous paralysis of adults at 29°C and larvae at 37°C. Recovery from paralysis at 25°C is immediate. A number of additional *para* alleles have since been isolated that fall into several different categories (48): recessive temperature-sensitive alleles similar to $para^{ts1}$ but usually requiring a higher temperature to cause paralysis; a dominant temperature-sensitive allele; null alleles that cause an unconditional recessive lethal phenotype; chromosome aberrations with one breakpoint in *para* that have the properties of *para* null alleles; recessive lethal and temperature-sensitive alleles associated with the insertion of transposable elements. The latter class of mutant alleles has been especially important for the cloning of *para* (see below).

The similarity in behavioral defects between nap^{ts} and $para^{ts1}$ is paralleled by a similarity in their electrophysiological phenotype; loss of the excitatory junctional potential occurs in $para^{ts}$ larvae at restrictive temperatures (9). Failure to evoke an excitatory junctional potential can be traced specifically to a block in the propagation of action potentials in the larval nerve bundle (9,48) (*Figure 5*). Similar results have been observed in electrophysiological studies of adults (11,13). The array of physiological phenotypes manifested by nap^{ts} at permissive temperatures are not observed in $para^{ts1}$ and toxin-binding assays have not revealed any obvious differences from wild-type. However, like nap^{ts}, some temperature-sensitive *para* alleles are able to suppress the hyperexcitable behavioral phenotype of *Sh* mutants at permissive temperatures (M.Stern and B.Ganetzky, unpublished data). In addition, cultures of $para^{ts1}$ neurons from dissociated larval ganglia are more resistant than normal to veratridine cytotoxicity, although less so than nap^{ts} neurons. At restrictive temperatures, veratridine resistance of $para^{ts1}$ neurons is comparable to that of nap^{ts} (105). These results and the interaction of *para* mutants with nap^{ts} in double mutant combinations (see below) led to the suggestion that *para*, like *nap*, is affecting sodium channels in some way (9,48).

Recent experiments using the whole-cell clamp technique to study cultured embryonic neurons have provided direct evidence that *para* affects sodium current in these cells (109). In cultures of normal embryos, sodium currents can be recorded in about 65% of the neurons examined. Three mutant alleles of *para* were examined and each caused a reduction in the percentage of cultured cells that expressed sodium currents. The magnitude of the reduction varied in an allele-dependent manner: $para^{ts1}$, $para^{ts2}$ and $para^{ST76}$ displayed 22, 34 and 95% reductions, respectively (109). It is not yet known whether the current kinetics and voltage-dependence are normal in those *para* cells that do express sodium currents.

Molecular evidence from the cloning of *para* provides very strong evidence for a direct relationship with sodium channels. The *para* locus was cloned by means of transposon tagging (48). Among 10^5 progeny from dysgenic crosses in which the P element was mobilized, two newly arisen *para* mutants were recovered. *In-situ* hybridization demonstrated that one of these mutations, $para^{hd2}$, was associated with a P-element insertion at 14C6-8, the exact chromosomal location where *para* had been previously localized by cytogenetic techniques. Furthermore, when P elements were remobilized in $para^{hd2}$, revertants or back mutations to $para^+$ were recovered. These reversion events were correlated with loss of P-element DNA from the insertion site in the *para* region. These data argued conclusively that insertion of a P element in or very close to $para^+$ was the cause of the $para^{hd2}$ mutant phenotype (48).

A genomic phage library was constructed from a $para^{hd2}$ strain in which most of the extraneous P elements had been removed by recombination and screened for clones that hybridized to a P-element probe. To identify those cloned genomic DNA fragments that were derived from the *para* locus rather than from another site of P-element insertion, each isolated DNA segment was analyzed by *in-situ* hybridization to polytene chromosomes from a strain lacking P elements. Cloned DNA segments from the *para* region were thus distinguished by their hybridization to region 14C6-8. One of these DNA segments was used to initiate a chromosome walk of over 200 kb in the *para* region (48; K.Loughney and B.Ganetzky, unpublished data). A number of mutant *para* alleles associated with chromosome breakpoints or insertion of transposable elements have been located on the molecular map of the region. The region spanned by the various *para* mutants extends over 45 kb, providing a minimum size estimate of the region essential for normal *para* function.

Probes from this 45 kb interval of genomic DNA were used to screen cDNA libraries to determine which portions were transcribed into mRNA. Several incomplete cDNA clones have been recovered that map to those regions of the walk where various *para* alleles have been located (K. Loughney and B.Ganetzky, unpublished data). It is believed that these cDNA clones represent pieces from a single large transcript. DNA sequence analysis of two of these clones revealed that each contains an open reading frame capable of encoding an amino acid sequence closely related to that of the vertebrate sodium channel. One clone encodes a stretch of 99 amino acids that can be aligned exactly with a portion of the third homologous repeat of the rat channel that includes the S4 and S5 domains and the region between them. In this region the *Drosophila* and rat amino acid sequences are 53% identical. If conserved amino acid changes are taken into consideration, the conservation between the two sequences is 62%. In the S4 region the positions of the charged amino acids is completely conserved. The second clone encodes a stretch of 87 amino acids that can be similarly aligned with a portion of the fourth homologous repeat of the rat channel that includes the membrane-spanning regions S1, S2, a portion of S3, and the regions between them. In this region the *Drosophila* and rat sequences are 49% identical and 68% conserved. Homology values similar to these are obtained in comparisons of the corresponding regions in the gene cloned by Salkoff *et al.* (34; see above) with the rat sodium channel. Comparison of the amino acid sequences for the two *Drosophila* genes also indicates about 50% identity.

The degree to which *para* and the rat sodium channel gene(s) are related is apparent from a different set of experiments that also resulted in the isolation of clones of genomic DNA from the *para* locus. By using a DNA fragment from the cloned rat sodium channel gene to screen *Drosophila* genomic libraries at low stringency, several clones that map within the boundaries of the *para* region have been isolated (M.Ramiswami and M.Tanouye, personal communication).

The above molecular analysis demonstrates that the *para* encodes a protein that shares regions of extensive amino acid similarity with the α subunit of vertebrate sodium channels. This result together with the physiological defects seen in *para* mutations strongly suggest that the *para* product represents at least one type of *Drosophila* sodium channel. If this suggestion proves correct it will be of interest to determine how the expression and distribution of the *para* gene product compares with that of the putative sodium channel gene identifed by Salkoff *et al.* (34).

4.2.3 The tip-E locus

The *tip-E* (temperature-induced paralysis E) locus is defined by a recessive temperature-sensitive paralytic mutation on the third chromosome (110). Behaviorally, *tip-E* resembles nap^{ts} and $para^{ts1}$, although paralysis of adults and larvae does not occur at temperatures less than 39–40°C (10,110,111). Action potentials in *tip-E* larval nerves are not blocked at temperatures up to 40°C (10). However, in double mutant combinations with *nap* or *para* alleles (see Section 4.3), *tip-E* manifests more striking effects on viability, behavior and membrane excitability suggesting that it has some functional overlap with *nap* and *para* (10,111). Consistent with the idea that *tip-E* affects sodium channels is its effect on STX binding: head-membrane preparations from *tip-E* have about a 40% reduction in the total number of binding sites but the affinity of binding and the pH profile of binding are normal (111). The reduction in toxin binding is more apparent at 38°C than at 0°C, suggesting that the mutant toxin-binding sites are abnormally labile. From these results it was postulated that *tip-E* alters either the the structure of sodium channels or their microenvironment (111). That *tip-E* alters sodium channels in some way is borne out by the observation that sodium current density in cultured embryonic neurons is reduced by 30–40% (D.O'Dowd, S.Germeraad, and R.Aldrich, personal communication). Whether this reduction reflects a change in channel number of single-channel conductance in the mutant neurons awaits further analysis by single-channel experiments.

X-ray-induced deletions and translocations mutant for *tip-E* have recently been generated enabling the locus to be mapped cytologically (L.M.Hall, D.W.Gil, and D.P.Kasbekar, personal communication). A chromosome walk in the *tip-E* region as defined by the deletions and translocations has also been initiated to obtain a molecular description of the *tip-E* product.

4.2.4 The sei locus

Mutations at a fourth locus, *sei* (seizure), implicated in sodium channel structure and function were also identified by the phenotype of temperature-sensitive paralysis but the behavioral phenotype is distinct from that of the other mutants in this category. Homozygotes for either $sei^{ts1}2$ or sei^{ts2} cause paralysis of adults but not larvae at the restrictive temperature of 38°C (B.Ganetzky and C.-F.Wu, unpublished data; 102). Mutant adults do not paralyze instantly but undergo a severe second bout of uncontrolled flight activity when exposed to 38°C. Thereafter, the flies fall down and are unable to right themselves but continue to show considerable movement of their legs. Recovery of sei mutants from paralysis at 25°C is not immediate but varies inversely with the length of prior exposure to the restrictive temperature (102,103).

The behavioral phenotypes of sei^{ts1} and sei^{ts2}, the two extant alleles, are similar but the alleles differ from each other in other respects (102,103). The sei^{ts2} allele is co-dominant with sei^+, whereas sei^{ts1} is recessive. In addition, the two alleles cause distinctive alterations in toxin-binding properties. Flies heterozygous for sei^{ts2} become paralysed at 38°C with kinetics intermediate between that of wild-type and sei^{ts2} homozygotes. Using a duplication that carries sei^+, it is possible to construct flies that carry three doses of the sei locus and to vary the relative dosage of sei^{ts2} and sei^+. Flies carrying a single copy of sei^{ts2} still become paralyzed even if they also carry two copies of the wild-type sei allele. Kinetics of paralysis at the restrictive temperature becomes increasingly more rapid as the dosage of sei^{ts2} increases relative to sei^+. These results argue that the altered sei^{ts2} product actively interferes with the normal function of the wild-type product. Ligand-binding studies also point to a structurally altered gene product. Membrane extracts from the heads of sei^{ts2} flies have a normal number of STX-binding sites and normal affinity of binding at 4°C. However, at 39°C the affinity of binding in sei^{ts2} extracts differs from normal by 2-fold while the number of binding sites remains normal (102). The pH profile of STX binding at 39°C also differs from normal indicating a possible alteration in the pKa of the STX-binding site. There is a dose-dependent effect of sei^{ts2} on ligand binding that parallels its effect on behavior; as the dosage of sei^{ts2} increases relative to sei^+ there is a proportional increase in the level of the low affinity binding of STX (103). This outcome is the result that would be expected if the amount of the lower affinity sei^{ts2} product in membrane extracts is increased relative to the normal product.

In contrast to these observations, sei^{ts1} acts as a fully recessive mutation; a single dose of sei^+ is sufficient to prevent the paralysis of flies carrying two doses of sei^{ts1} (103). In toxin-binding assays, the affinity of sei^{ts1} extracts for STX is normal at both 0 and 39°C. However, the total number of binding sites is reduced by 5–18% at 0°C and 17–37% at 39°C. The pH dependence of STX binding is normal in sei^{ts1} extracts (103). From the results of toxin-binding studies it was proposed that sei encodes a structural component of sodium channels. The differences between sei^{ts1} and sei^{ts2} could be explained if they affect different structural domains of the polypeptide with sei^{ts2} altering the properties of the binding site and sei^{ts1} directing the synthesis of a product with increased lability.

Alterations in sodium current in sei^{ts2} neurons have not yet been reported but a reduction of about 30–40% in sodium current density relative to wild-type was found in cultured embryonic sei^{ts1} neurons (D.O'Dowd, S.Germeraad, and R.Aldrich, personal communication). Although the latter result is consistent with the conclusion that sei^{ts1} directly causes the production of unstable sodium channels, there are still some problems with this interpretation. The only electrophysiological abnormality that has been observed in sei^{ts1} and sei^{ts2} adults is an apparent increase in spontaneous nerve activity at 40°C in the GFP (112,113). This spontaneous activity could be an electrophysiological correlate of the behavior displayed by sei mutants at the restrictive temperature. However, the result is surprising for two reasons:

(i) if sei^{ts1} causes a reduction in sodium current density, it seems more likely that it would decrease rather than increase nerve activity;
(ii) if, as proposed, sei^{ts1} and sei^{ts2} are altering sodium channels in very different ways their similar effects on adult nerve activity would not be expected.

It is unlikely that these issues will be settled until the sei locus has been cloned and its gene product identified. At present, direct molecular evidence that sei encodes a structural component of sodium channels is lacking.

4.3 Double mutant interactions

The analysis of various double mutant combinations led to the discovery of striking synergistic interactions involving some of the mutations affecting sodium channels arguing for significant functional overlaps or interactions among their gene products. For example, $para^{ts1}$;nap^{ts} double mutants die at some point during larval development regardless of the temperature at which they are raised (9,114). In viable mosaics, bristle sensory neurons located in small patches of doubly mutant cuticle differentiate normally but fail to propagate action potentials at all temperatures (115). It is this unconditional block in nerve conduction that is believed to be responsible for lethality of the $para^{ts1}$;nap^{ts} double mutant.

All temperature-sensitive alleles of para examined were found to interact synergistically with nap^{ts} but the strength of the interaction varies in an allele-dependent manner (114). With alleles other than $para^{ts1}$, varying percentages of double mutants survive through eclosion. However, in all cases the survivors are weak and become paralyzed at temperatures only one or two degrees above 25°C, the temperature at which they are raised. The allele-specific interactions suggest the possibility of a direct physical interaction between the *nap* and *para* gene products (114).

It seems likely that even the normal function of *para*$^+$ depends on *nap*$^+$ as evidenced by the dominant effects exhibited by mutant *para* alleles in a *nap*^{ts} background (114). For example, in a *nap*$^+$ background, heterozygotes for *para*^{ts1} and *para*$^+$ have normal viability and they do not show temperature-sensitive paralysis. The same heterozygote in a homozygous *nap*^{ts} background has significantly reduced viability and the restrictive temperature for paralysis of the survivors is very low (27°C). Heterozygotes for a recessive lethal *para* allele, which are viable in a *nap*$^+$ background, are completely non-viable in a homozygous *nap*^{ts} background.

Similar results have been observed in double mutants of *para* with *tip-E*. In combination with *tip-E*, different *para* alleles have greatly reduced viability through eclosion and in the most extreme cases are completely non-viable (10,111). Double mutants for those *para* alleles able to survive to some degree in combination with *tip-E* are very inactive, become paralyzed, suffer a block in nerve conduction at temperatures where the single *para* or *tip-E* mutations would not manifest these phenotypes, and die within a few days after eclosion. Surprisingly, the *para* alleles that interact strongest with *nap*^{ts} are those that interact the weakest with *tip-E* and vice versa. That *para* interacts with both *tip-E* and *nap*^{ts} in an allele-dependent fashion but with a different ordering of alleles in each case, argues that it is not simply the residual level of wild-type activity in a particular *para* allele that governs the severity of its interaction with *nap*^{ts} or *tip-E*. Instead, the idea of a direct physical interaction with the *nap*^{ts} and *tip-E* products recognizing certain aspects of the *para* product has been suggested (10). If, as the previously described molecular analysis is beginning to indicate, the *para* locus encodes a structural component of sodium channels, the molecular identification of the *nap* and *tip-E* gene products and the elucidation of their interaction with the *para* product may shed new light on mechanisms that regulate the production of sodium channels or modulate their activity.

The functional relationship between *nap*^{ts} and *tip-E* is evident not only from their similar interactions with *para* but from their interactions with each other. Although some *nap*^{ts};*tip-E* double mutants survive to eclosion, they are much more sensitive to temperature-induced paralysis and nerve blocks than either single mutant, and die within several days after eclosion (10,111). The effects of *nap*^{ts} and *tip-E* on STX binding are cumulative so that the reduction in STX binding sites measured in head-membrane extracts from *nap*^{ts};*tip-E* double mutants (60–65%) is significantly greater than in extracts from either *nap*^{ts} or *tip-E* alone (111). This result suggests that *nap*^{ts} and *tip-E* do not cause a reduction in toxin-binding via the identical mechanism.

Synergistic interactions of the type above were not observed in double mutant combinations with *sei*^{ts2}. The most noticeable effect in double mutants of sei^{ts2} with nap^{ts} or tip-E was the accelerated recovery from paralysis relative to sei^{ts2} alone following exposure to the restrictive temperature (103). This result was interpreted as evidence that enhanced neuronal excitability at the restrictive temperature is responsible for the sei^{ts2} behavioral phenotype and this increased nerve activity is blocked or reduced by nap^{ts} or tip-E. This interpretation is consistent with the identification of another mutation, enhancer of seizure [$e(sei)$], which lowers the restrictive temperature of sei^{ts1} and sei^{ts2} and also causes higher rates of spontaneous nerve activity in the GFP circuit at restrictive temperatures in double mutants (112).

5. Molecular and genetic analysis of potassium channels

In *Drosophila*, as in other organisms, a variety of potassium channels with diverse biophysical properties are known and are distinguishable by pharmacological as well as physiological criteria (cf. refs 98,116) (*Figure 6*). At least four distinct types of potassium channel have been examined in *Drosophila* (14,18,20,117–120) that separately mediate an early and a late voltage-dependent current (I_A and I_K) and an early and a late calcium-dependent current (I_C and I_{KC}). A characteristic of both early currents is that they are transient owing to the self-inactivation of the channels mediating these currents shortly after the channels open. In contrast, the delayed currents show little inactivation even during sustained depolarization or calcium influx. The molecular basis for the difference in these biophysical properties among potassium channels remains to be elucidated.

A variety of potassium channels has presumably evolved to mediate specialized functions within different cell types or cellular regions (cf. ref. 98). On the basis of recent molecular evidence demonstrating evolutionary relationships among different classes of ion channels (121), it is reasonable to propose that the distinct potassium channel types now in existence arose from a common ancestor. However, this supposition remains unproven because of the present scarcity of molecular genetic analyses of potassium channels. In contrast to sodium channels, a potassium channel polypeptide has not yet been purified to homogeneity from any organism, precluding the cloning of the corresponding genes by use of oligonucleotide probes. Thus, the strategy that has worked so well for the cloning of other ion channel genes may not prove as applicable for potassium channels.

The ability to isolate mutations affecting potassium channels in *Drosophila*, some of which may reside in structural genes, permits an alternative strategy for molecular analysis that does not require the prior purification of a potassium channel polypeptide. Mutations of several different genes detected initially because of behavioral defects have been shown by direct measurements to alter different potassium currents. As previously described, once a mutation is identified and its cytological location known, methods are available in *Drosophila* to clone the gene without any biochemical information about the encoded product (see Section 1.3.2). The power of this approach is exemplified by the electrophysiological and genetic analysis of the *Sh* (*Shaker*) locus showing that is is a likely structural gene for I_A channels, followed by the cloning

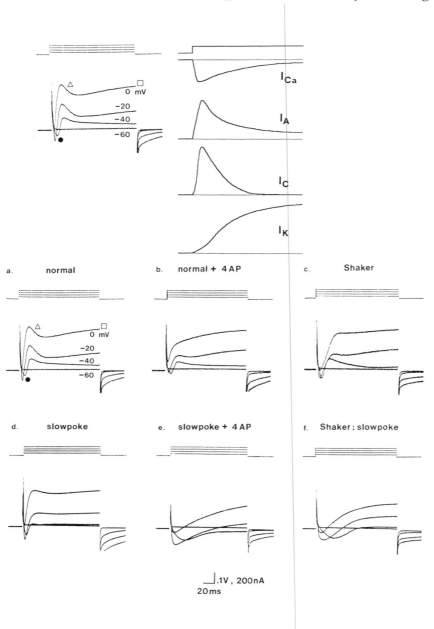

and molecular analysis of this gene (see below). These studies have provided the first molecular description of a putative potassium channel polypeptide from any organism. Mutations affecting other classes of potassium channels are now being subjected to similar molecular analysis. Ultimately, this approach should help elucidate the evolutionary relationships among different types of potassium channels and the structural basis for their distinctive biophysical properties.

5.1 Mutations affecting potassium channels

5.1.1 The Sh locus

Mutations of the X-linked *Sh* locus were discovered by their phenotype of aberrant leg shaking under ether anesthesia (122). In *Sh* adults, action potentials in the giant axon of the GFP display delayed repolarization often coupled with repetitive firing (3,4). In *Sh* larvae, release of neurotransmitter at neuromuscular junctions is greatly enhanced and prolonged and is correlated with abnormal repetitive firing of motor axons (7,8,123) (*Figure 7*).

Voltage-clamp measurements of muscle membrane currents in embryonic cell cultures, larvae, pupae and adults demonstrate that *Sh* mutations specifically affect I_A (14,17–20,26,117) (*Figure 6*). Several *Sh* alleles abolish I_A completely. Other alleles reduce the amplitude of I_A but retain its normal kinetic properties, consistent with a reduction in the number of functional channels. Only a single allele has been identified that alters kinetic properties of I_A channels; Sh^5 alters the rate and voltage dependence of inactivation of I_A in DLMs (124). In larval muscles the voltage dependence of activation as well as inactivation is altered and the amplitude of I_A is reduced (125). From the electrophysiological

Figure 6. Membrane currents in adult dorsal longitudinal flight muscles (DLMs). Top left: Current responses (lower traces) to voltage-clamp steps (upper traces) in a normal DLM fiber. The membrane was held at −80 mV stepped to −60, −40, −20, and 0 mV for 140 ms, as indicated. The pulse to −60 mV activates only passive leakage current. At −40 mV an inward (downward) Ca^{2+} current (closed circle) is followed by an inactivating outward current (triangle). This outward current is separable into two components, the voltage-dependent I_A and the Ca^{2+}-dependent I_C. At −20 mV, a delayed outward current, I_K, is evident (square). (Top right) Idealized diagram of the individual ion currents, showing their approximate amplitudes and time courses in response to the voltage step to 0 mV above. Bottom: Comparison of current responses to voltage-clamp steps in normal, mutant and drug-treated DLMs. (**a**) Normal; (**b**) normal treated with 5 mM 4-aminopyridine, which selectively blocks I_A (**c**) shaker mutant (Sh^{KS133}), which lacks I_A; (**d**) slowpoke (*slo*) mutant, which lacks I_C; (**e**) slowpoke mutant treated with 5 mM 4-aminopyridine; (**f**) Sh^{KS133}; *slo* double mutant. Calibration: 0.1 V, 200 nA, 20 ms; Temperature, 4 °C. Top figures are reproduced from ref. 134; bottom figures are reproduced from ref. 14.

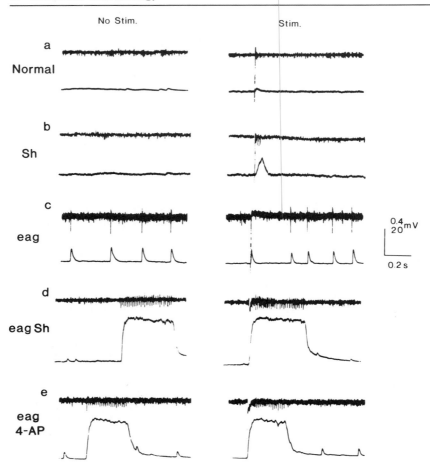

Figure 7. Simultaneous nerve and muscle recordings from normal, Shaker (*Sh*), ether a go-go (*eag*) and *eag Sh* larvae. Excitatory junctional potentials were recorded intracellularly from body-wall muscle fibers and the motor axon activities were recorded via a suction electrode. Segmental nerves were cut close to the ganglion. Activities evoked by nerve stimulation (right panels) and occurring spontaneously without stimulation (left panels) are shown. The motor axon spikes are the most prominent units but their absolute amplitude can vary depending on recording conditions. At low external Ca^{2+} concentrations [0.2 mM for (**a**) and (**c**) and 0.1 mM for (**b**), (**d**) and (**e**)] the nerve stimulation evokes, enhanced neuromuscular transmission in mutant (**b**–**e**) as compared with normal (**a**) larvae. In *Sh* a single nerve stimulus causes the motor axon to fire several extra spikes causing enhanced release of transmitter (**b**). Slow but rhythmic repetitive firing of motor axons, correlated with junctional potentials, occurs spontaneously in *eag* (**c**) but not in normal (**a**) or *Sh* (**b**) larvae. Combining *eag* and *Sh* in double mutants produces striking synergistic effects, resulting in greatly prolonged junctional potentials correlated with bursts of motor axon spikes (**d**). These events occur in response to nerve stimulation or spontaneously without stimulation. This phenotype of the *eag Sh* double mutant is exactly mimicked by treating *eag* larvae with 4-AP. Reproduced from ref. 132.

phenotypes displayed by the various *Sh* alleles individually and in hteroallelic combinations, it has been proposed that the *Sh* locus encodes a structural component(s) of I_A channels (for detailed reviews see refs 52,124–127). Analysis of segmental aneuploids for the *Sh* locus supports the idea that it is a structural gene for I_A channels; deletions of the *Sh* locus result in complete loss of I_A while addition of an extra copy of the *Sh* locus by means of a duplication produces a proportional increase in I_A, at least in larval muscles (125).

The existence of chromosome translocations associated with *Sh* behavioral and electrophysiological phenotypes enabled the *Sh* locus to be mapped cytologically to region 16F on the X chromosome (4). Fortuitously, this location is very close to the site where cDNA clone *adm135H4*, previously isolated for purposes entirely unrelated to *Sh*, mapped by *in-situ* hybridization. Several laboratories took advantage of this proximity to carry out a chromosome walk through the *Sh* region using *adm135H4* as a starting point (43–45). The most distal translocation shown to affect *Sh* is *T(1;Y)B55*. The most proximal translocation that clearly affects *Sh* is *T(1;Y)W32*. These translocations define the minimum extent of the *Sh* region and provide physical landmarks, recognizable by *in-situ* hybridization and Southern blot analysis, enabling the localization of the *Sh* locus within the cloned region. The breakpoints of *T(1;Y)B55* and *T(1;Y)W32* were identified on the molecular map of the walk and were found to be separated by about 65 kb. It was expected that most (but not necessarily all) of the relevant coding sequences of the *Sh* gene would be located in this interval. Subcloned fragments from this region of the walk were used to screen cDNA libraries to identify transcribed regions. A collection of cDNA clones that derive from the *Sh* region have been isolated in this way (43–45). The cDNA clones contain exons that span more than 60 kb of genomic DNA [95 kb in the case of a cDNA clone reported by Kamb et al. (43)] and encompass the mutant sites associated with various mutant *Sh* alleles including several translocations, an inversion, and an insertional mutation. Although the cDNA clones differ from each other, they fall into several discrete categories and all share exons in common. These results indicate that the *Sh* locus encodes a family of transcripts that differ from one another by differential splicing.

The nucleotide sequence determined for several of the cDNA clones provides evidence that *Sh* encodes a structural component of potassium channels (43,45,128). The most complete cDNA analyzed thus far encodes a polypeptide of 616 amino acids (128). Hydropathy analysis of this polypeptide indicates that it has a central hydrophobic core with hydrophilic domains at its amino and carboxyl termini. Seven potential membrane-spanning regions were identified indicating that the *Sh* product is an integral membrane protein. Evidence that it is a channel protein derives from comparison of the *Sh* amino acid sequence with that of the eel sodium channel. A single region of amino acid similarity is found for these two proteins; over a stretch of 121 amino acids the proteins are 27% identical and 47% conserved. The region of similarity centers on the arginine-rich S4 domain of the eel protein, whose proposed involvement with the mechanism of voltage-dependent gating of the sodium channel was described earlier (see Section 4.1). The number and spacing of the charged amino acids is completely conserved in this region between the *Sh* polypeptide and the S4 domains of the fourth homology units of the rat and the eel (*Figure 4*). It therefore seems very likely that an S4-like domain plays a functionally important role in the gating of the potassium channels specified by the *Sh* locus. Furthermore, the presence of an S4 domain in potassium channels, sodium channels (100,101) and calcium channels (129) suggests that the voltage-gating mechanism arose in evolution prior to the divergence of various species of voltage-activated channels and that this mechanism has been highly conserved since its origin. The amino acid sequence encoded by another *Sh* cDNA lacked the S4 domain but sequence similarities with the S1, S2 and S6 membrane-spanning domains of the vertebrate sodium channel were reported (45). As the first description in any organism of the amino acid composition and organization of a potassium channel polypeptide and the first time that a channel-encoding gene has been cloned prior to the purification of the relevant protein, the membrane analysis of *Sh* illustrates the power of a genetic approach to channel structure and function.

A number of interesting problems concerning the structure of I_A channels and the role of *Sh* remain to be addressed. Compared with the α subunit of sodium (100,101) or calcium channels (129), the *Sh* polypeptide is small and is analogous to just one of the four internally repeating units that constitute a sodium or calcium channel polypeptide. Consequently, it seems unlikely that the *Sh* polypeptide alone constitutes the complete I_A channel. Rather, these channels are probably multimeric structures. The question then remains whether polypeptides encoded by other genes are also structural components of I_A channels or if the channels contain multiple polypeptides all encoded at the *Sh* locus. In the latter case, I_A channels could be a homo-multimeric structure comprising an unknown number of identical *Sh* polypeptides. Alternatively, distinct polypeptides that could arise from the differentially spliced *Sh* transcripts could be part of the channel structure.

Molecular analysis of the different products specified by the *Sh* locus should shed new light on inferences about the composition of I_A channels based on voltage-clamp studies. Measurements of I_A in body wall muscle of larvae heterozygous for different combinations of *Sh* alleles indicated that in most cases the amplitude of the residual I_A is exactly midway between that of the corresponding homozygotes (127). The simplest interpretation of this result is that all of these *Sh* alleles affect the same gene product and that there is only one copy of this product per I_A channel. Heterozygous individuals would then have two equal populations of channels in muscles, each population specified by a particular allele. However, the results with some *Sh* alleles (Sh^{KS133} and Sh^{102}) did not fit this simple picture (127). Heterozygotes with these alleles had an I_A that was much less than expected from the average of the corresponding homozygotes. One possible explanation is that alleles such as Sh^{KS133} and Sh^{102} specify a subunit of I_A channels distinct from that affected by the other *Sh* mutants and that this subunit represents a multimeric component of the channels. In this case, the mutant phenotype could prevail if even a single mutant subunit is included in the multimer.

The generation of multiple gene production from the *Sh* locus also raises the question of whether the different polypeptides give rise to a family of I_A channel subtypes that are found in different cells, different cellular regions, or different developmental stages. The electrophysiological defects manifested by various *Sh* mutants indicate that the *Sh* locus must function in different excitable cells and at different developmental stages. However, analysis of particular *Sh* mutants indicates that the severity of the defect associated with these alleles varies depending on the cell type and developmental stage examined. For example, at the larval neuromuscular junction, the effects of Sh^{rKO120} and Sh^{KS133} on synaptic transmission are equally severe (130), whereas the effect of Sh^{KS133} on repolarization of the giant axon or on I_A in pupal and larval muscle is much greater than that of Sh^{rKO120} (3,4,20,124,131). Perhaps Sh^{rKO120} affects a product of the *Sh* locus primarily important for the function of I_A channels in pre-synaptic nerve terminals.

Similarly, recent patch-clamp studies demonstrate the presence of an I_A-type current in cultured embryonic muscle cells in addition to cultured neurons from the larval CNS (26). However, the detailed physiological properties of the I_A current recorded from these two cell types are quite distinct. Furthermore, Sh^{KS133} eliminates I_A in the embryonic muscle cells but does not affect I_A in the larval CNS cell bodies (26). Since Sh^{KS133} is known to affect excitability of neuronal axons and pre-synaptic terminals (3,4,7,8,123), these results suggest that the products of the *Sh* locus may not have exactly the same role in controlling I_A channels in different types of excitable cells or membrane regions.

In principle the molecular tools are now available to explore the kind of issues raised above. It will be of interest to learn whether the details of *Sh* expression and regulation will be generally applicable to other potassium channel genes in *Drosophila* and other organisms.

5.1.2 The eag locus

The *eag* (ether a-go-go) locus was identified on the basis of an X-linked mutation that caused ether-sensitive leg-shaking behavior (122). Subsequently, a mutation discovered as an enhancer of the *Sh* behavioral and electrophysiological mutant phenotypes was identified as an *eag* allele (130). Electrophysiological investigations showed that *eag* mutations cause the abnormal discharge of spontaneous repetitive action potentials in larval motor axons leading to a high frequency of spontaneous excitatory junctional potentials (130,132) (*Figure 7*).

Sh and *eag* interact synergistically in double mutants, producing extreme membrane hyperexcitability (130,132) (*Figure 7*). Release of neurotransmitter at the neuromuscular junction in double mutant larvae persists at least ten times longer than in either single mutant, generating a very large plateau-shaped excitatory junctional potential. These distinctive excitatory junctional potentials occur both in response to nerve stimulation and spontaneously without nerve stimulation and are correlated with the occurrence of long trains of action potentials in the motor axon. These prolonged bursts of high frequency spikes in the double mutant are quite distinct from the slow, pacemaker-like spontaneous activity seen in motor axons of *eag* alone. The synergistic interaction between *Sh* and *eag* in double mutants led to the proposal that *eag*, like *Sh*, was defective in repolarization of neuronal membranes.

Results of pharmacological studies are consistent with this interpretation since the phenotype of the double mutant can be mimicked by applying tetraethylammonium (TEA), which is known to block both I_A and I_K in *Drosophila* muscles (18,20), to the neuromuscular junction of normal larvae (123). Application of 4-aminopyridine (4-AP), which blocks I_A preferentially (18), also mimics the double mutant phenotype when applied to the neuromuscular junction of *eag* but not normal larvae (130,132). These results suggest that *eag* affected a potassium current in neurons that corresponded to the muscle I_K. Voltage-clamp studies of eag^1 larval muscles revealed that I_K was, in fact, significantly reduced in these cells (19).

The phenotype of other *eag* alleles at the voltage-clamp level leads to a more complicated picture. None of the *eag* alleles examined, including some thought to be null alleles, abolish I_K completely (Y.Zhong, R. Drysdale, C.-F.Wu and B.Ganetzky, unpublished data), suggesting that some other gene(s) must also play a major role in the production of I_K channels. In addition, some *eag* alleles (eag^{24}, eag^{4PM}) were found to alter I_A rather than I_K, while another allele (eag^{x6}) affects both I_A and I_K (Y.Zhong, R.Drysdale, C.-F.Wu and B.Ganetzky, unpublished data). These data suggest that *eag* may encode a component shared by different types of potassium channels or that it plays a role in modulating the activity of these channels.

To identify the *eag* gene product and to obtain better understanding of the *eag* mutant phenotype, a molecular analysis of the *eag* locus has been undertaken (46,133). Clones from the *eag* region were obtained by taking advantage of an *eag* mutation that is associated with a chromosome inversion broken at one end in or very close to the *eag* locus and broken at the other in the *sc* (scute) locus, which had previously been cloned. A junction fragment of DNA, which includes a portion of the *eag* region, was isolated by screening a genomic library of the inversion using available probes from the *sc* region. The *eag* portion of the junction fragment was then used to screen a wild-type library to isolate additional overlapping clones in the *eag* region. Over 60 kb of DNA from the *eag* region have now been cloned and mapped. The extent of the *eag* gene within the cloned region is being defined by locating on the map the alterations associated with different *eag* alleles (45). The molecular lesions of an insertional translocation (*eag*^{x6}) produced by X-irradiation and two insertional mutations produced by hybrid dysgenesis (*eag*^{hd14} and *eag*^{hd15}) have been mapped. The insert in *eag*^{hd15} was shown to be a P element by *in-situ* hybridization. Three revertants, or back mutations to wild-type, derived from *eag*^{hd15} after subjecting it again to dysgenic conditions, all lost the original insertion proving that the insertion was the cause of the mutant phenotype. The lesions associated with *eag*^{x6} and *eag*^{hd14} fall within 5 kb of *eag*^{hd15}. Thus, these three alleles define a region of the walk critical for normal *eag* function. However, at the present time it is not known which portions of the *eag* region contain coding information since the breakpoint of the inversion used to clone the region is located 20–25 kb from the *eag*^{hd15} region. Experiments to identify the *eag* transcript on Northern blots and to isolate corresponding cDNA clones are now in progress. From the nucleotide sequence of cDNA clones it should be possible to distinguish whether *eag* encodes a structural component of potassium channels or a protein involved in their regulation.

5.1.3 The *Hk* locus

The *Hk* (Hyperkinetic) locus is another X-linked gene, mutations of which cause ether-induced leg-shaking behavior (122). Early investigations of *Hk* indicated abnormal membrane properties in certain groups of neurons in defined regions of the adult thoracic ganglion (1,2). Intracellular recordings from these cells revealed the abnormal endogenous production of action potentials with a rhythm that corresponded to that of the leg-shaking activity.

More recently, patch-clamp studies of cultured larval CNS neurons indicated that a type of potassium channel that mediates an inward rectification current (which is activated by hyperpolarizing the membrane) has an unusually large unit conductance in *Hk* (23). Because potassium ions flow inward through these channels until the membrane potential is about 20 mV greater than the resting potential, activation of these high conductance channels could drive the membrane potential towards the action potential threshold. This might be a contributing factor in the endogenous activity of the *Hk* neurons described above.

Additional studies indicate that the effect of *Hk*, like *eag*, is not limited to a single type of potassium channel. In muscle fibers of normal *Drosophila* adults and larvae the inward rectification current seen in neurons is not found. However, *Hk* appears to alter I_A in these cells; the kinetics of I_A activation is slower than normal and the peak amplitude of the current is reduced (C.-F.Wu and T.Elkins, unpublished data). *Hk* larvae display enhanced release of transmitter at the neuromuscular junction associated with repetitive firing of the motor axon following a bout of high frequency nerve stimulation, which may also be explained by a deficit in I_A (134). Further genetic studies of the *Hk* locus are now in progress to facilitate the molecular isolation and characterization of this gene (A.K.Schlimgen and B.Ganetzky, unpublished data).

5.1.4 *The slo locus*

Voltage-clamp analysis of DLMs in Sh^{KS133} adults in calcium-containing saline revealed that the early, inactivating outward current observed in wild-type flies had a second component distinct from I_A (117). This component is not blocked by 4-AP but is eliminated by charybdotoxin and in calcium-free Ringer's solution (14,18,117). The second component was identified as a calcium-dependent potassium current (denoted I_C) similar to that found in other systems (98,116) (*Figure 6*). Despite the similarity in amplitude and kinetics of I_A and I_C, *Sh* mutations that completely eliminated I_A have no effect on I_C demonstrating that the channels mediating these currents are under separate genetic control (14,117). This conclusion was reinforced by the discovery of a mutation on the third chromosome, *slo* (slowpoke), that specifically eliminates I_C while leaving I_A unaltered (14). The *slo* mutant was detected in a screen for temperature-sensitive paralytic mutants because at 38°C *slo* flies are uncoordinated and unable to climb. Following exposure to 38°C for several minutes, *slo* flies remain standing motionless for a period of several minutes when returned to 22°C. Even without prior exposure to 38°C *slo* flies can be distinguished from wild-type by their poor flight ability and their ether-sensitive leg-shaking behavior.

Intracellular recordings of action potentials in DLMs evoked by stimulation of the GFP or by direct injection of depolarizing current revealed that the muscle spikes were abnormally prolonged in *slo*, persisting for about 20 ms rather than the usual 2 ms (14,135). Voltage-clamp experiments indicated that I_C was completely eliminated and that the early outward current in *slo* DLMs consisted of a single, 4-AP-sensitive component corresponding to I_A (14). This interpretation was confirmed by the voltage-clamp analysis of Sh^{KS133};*slo* double mutants, in which no early outward current is evident.

In larval muscles, an early calcium-dependent potassium current has also been studied (119,120). The channels that mediate this larval current share some molecular identity with the adult I_C channels because this current is also abolished by the *slo* mutation (120). In contrast, a late non-inactivating calcium-dependent potassium current (I_{KC}) has been detected in adult and larval muscles (118–120) and these currents are unaltered by *slo* (120). Thus, *slo* is defective in some structural component or regulatory mechanism that distinguishes the early and late calcium-dependent potassium channels.

The question of whether *slo* encodes a component of I_C channels is still open. In *Paramecium* a mutation that eliminates a calcium-dependent potassium current has been shown to reside in the structural gene for calmodulin (136). The same cannot be true for *slo* since the *Drosophila* calmodulin gene is located on a different chromosome. However, the possibility that *slo* alters some other regulatory protein cannot be ruled out. In the analysis of *Sh*, the observation that different alleles altered the I_A current in different ways provided an important clue that *Sh* was a likely structural gene for I_A channels. Similarly, if *slo* encodes a component of I_C channels it might be expected that certain *slo* alleles would reduce the density or modify the kinetics or calcium sensitivity of I_C rather than eliminate it entirely. New mutations are being isolated at the *slo* locus to address this issue (N.Atkinson and B.Ganetzky, unpublished data). Preliminary evidence indicates that some of the newly isolated alleles reduce but do not eliminate I_C (G.Robertson, N.Atkinson and B.Ganetzky, unpublished data).

Isolation of new radiation-induced alleles of *slo* associated with chromosome rearrangements will also enable the location of the gene to be pinpointed cytologically. Using procedures that have been previously outlined, it will then be possible to clone the gene for molecular analysis. If the *slo* gene product proves to be an I_c channel component, it will be informative to compare its structure with that of the *Sh* product since I_A and I_C channels both mediate a potassium current with similar kinetics and amplitude but differ in their mechanism of gating.

5.2 Interaction between sodium and potassium channel mutations

Sodium and potassium channels mediate currents that have opposing effects on neuronal excitation. Therefore, it might be expected that the phenotypic defects in mutants that depress the function of sodium channels could counterbalance some of the phenotypic defects in potassium channel mutants. This idea can be tested directly in appropriate double mutant combinations. For example, in double mutant combinations with *Sh* mutations, it has been found that nap^{ts} suppresses the leg-shaking behavior and aberrant activity of motor axons normally associated with these *Sh* alleles (8,104). These suppressive effects of nap^{ts} are observed at permissive temperatures. Apparently, the reduction in membrane excitability caused by nap^{ts} even at nominally permissive temperatures (see Section 4.2.1) is sufficient to counterbalance the hyperexcitable effect of reduced I_A caused by *Sh*. Similarly, nap^{ts} is able to suppress the leg-shaking phenotype of other mutations that affect potassium channels and cause membrane hyperexcitability such as *eag* and *Hk* (104).

The ability to suppress the phenotypes associated with various potassium channel mutations is not an exclusive property of nap^{ts}; new temperature-sensitive *para* alleles have recently been isolated that suppress the leg-shaking behavior and repetitive firing of motor axons of *Sh* mutations (M.Stern and B.Ganetzky, unpublished data). The same *para* alleles have been found to suppress the leg-shaking phenotype of *Hk* as well. As for nap^{ts}, suppression by the new *para* alleles occurs at nominally permissive temperatures. These results support the conclusion that at least some *para* alleles are similar to nap^{ts} in their ability to depress nerve membrane excitability even under permissive conditions.

6. Conclusions

Among metazoan organisms, *Drosophila* is perhaps unique in that sophisticated genetic, molecular and electrophysiological techniques can all be readily applied. This situation has permitted isolation of a wealth of mutations that affect signaling mechanisms in the nervous system at all levels from sensory input to motor output. These mutations provide an unparalleled set of tools for dissecting the process involved in neural signaling. Furthermore, such mutations can lead to the discovery of unsuspected roles for previously identified proteins as well as the discovery of new proteins. Application of recombinant DNA methodology is just beginning to give us the first detailed look at the proteins encoded by the genes thus identified and further rapid progress in this area can be expected. The use of *in-vitro* mutagenesis together with germ line transformation permits correlation of structural analysis of ion channels and related proteins with functional analysis *in situ*. This experimental system has significant advantages as compared, for example, with studies of cloned ion channel genes expressed in frog oocytes (e.g. 137). It therefore seems reasonable to anticipate that the multidisciplinary analysis to which *Drosophila* is now being subjected will yield novel insights about the molecular mechanisms of neuronal activity.

7. Acknowledgements

This article reviews work appearing in the literature or available to us as preprints by October 1987. We gratefully acknowledge many colleagues who sent us reprints and preprints and communicated to us their unpublished results for this article. Research in the authors' laboratories was supported by grants NS18500, NS15350 and Research Career Development Award NS00675 to C.-F.Wu, and by grants NS15390, GM35099 and Research Career Development Award NS00719 to B.

Ganetzky, all from the United States Public Health Service. B.Ganetzky is also a recipient of a Klingenstein Fellowship in the Neurosciences. Paper no. 2963 from the Laboratory of Genetics, University of Wisconsin, Madison.

8. References

1. Ikeda,K. and Kaplan,W.D. (1970) Patterned neural activity of a mutant *Drosophila melanogaster*. *Proc. Natl. Acad. Sci. USA*, **66**, 765–772.
2. Ikeda,K. and Kaplan,W.D. (1974) Neurophysiological genetics in *Drosophila melanogaster*. *Am. Zool.*, **114**, 1055–1066.
3. Tanouye,M.A. and Ferrus,A. (1985) Action potentials in normal and Shaker mutant *Drosophila*. *J. Neurogenet.*, **2**, 253–271.
4. Tanouye,M.A., Ferrus,A. and Fujita,S. (1981) Abnormal action potentials associated with the Shaker complex locus of *Drosophila*. *Proc. Natl. Acad. Sci. USA*, **78**, 6548–6552.
5. Tanouye,M.A. and Wyman,R.J. (1980) Motor outputs of giant nerve fiber in *Drosophila*. *J. Neurophysiol.*, **44**, 405–421.
6. Wu,C.-F., Ganetzky,B., Jan,Y.N., Jan,L.Y. and Benzer,S. (1978) A *Drosophila* mutant with a temperature-sensitive block in nerve conduction. *Proc. Natl. Acad. Sci. USA*, **75**, 4047–4051.
7. Jan,Y.N. and Jan,L.Y. (1980) Genetic dissection of synaptic transmission in *Drosophila melanogaster*. In *Insect Neurobiology and Pesticide Action*. F.E.Rickett (ed.), The Society of Chemical Industry, London, pp. 161–168.
8. Ganetzky,B. and Wu,C.-F. (1982) *Drosophila* mutants with opposing effects on nerve excitability: genetic and spatial interactions in repetitive firing. *J. Neurophysiol.*, **47**, 501–514.
9. Wu,C.-F. and Ganetzky,B. (1980) Genetic alteration of nerve membrane excitability in temperature-sensitive paralytic mutants of *Drosophila melanogaster*. *Nature*, **286**, 814–816.
10. Ganetzky,B. (1986) Neurogenetic analysis of *Drosophila* mutations affecting sodium channels: synergistic effects on viability and nerve conduction in double mutants involving *tip-E*. *J. Neurogenet.*, **3**, 19–31.
11. Siddiqi,O. and Benzer,S. (1976) Neurophysiological defects in temperature-sensitive mutants of *Drosophila melanogaster*. *Proc. Natl. Acad. Sci. USA*, **73**, 3253–3257.
12. Ikeda,K., Ozawa,S. and Hagiwara,S. (1976) Synaptic transmission reversibly conditioned by single-gene mutation in *Drosophila melanogaster*. *Nature*, **259**, 489–491.
13. Benshalom,G. and Dagan,D. (1981) Electrophysiological analysis of the temperature-sensitive paralytic *Drosophila* mutant *para*[ts]. *J. Comp. Physiol.*, **144**, 409–417.
14. Elkins,T., Ganetzky,B. and Wu,C.-F. (1986) A *Drosophila* mutation that eliminates a calcium-dependent potassium current. *Proc. Natl. Acad. Sci. USA*, **83**, 8415–8419.
15. Jan,L.Y. and Jan,Y.N. (1976) Properties of the larval neuromuscular junction in *Drosophila melanogaster*. *J. Physiol.*, **262**, 189–214.
16. Jan,L.Y. and Jan,Y.N. (1976) L-Glutamate as an excitatory neurotransmitter at the *Drosophila* larval neuromuscular junction. *J. Physiol.*, **262**, 215–236.
17. Salkoff,L. and Wyman,R. (1981) Genetic modification of potassium channels in *Drosophila* Shaker mutants. *Nature*, **293**, 228–230.
18. Salkoff,L. and Wyman,R.J. (1983) Ion currents in *Drosophila* flight muscles. *J. Physiol.*, **337**, 687–708.
19. Wu,C.-F., Ganetzky,B., Haugland,F.N. and Liu,A.-X. (1983) Potassium currents in *Drosophila*: different components affected by mutations in two genes. *Science*, **220**, 1076–1078.
20. Wu,C.-F. and Haugland,F.N. (1985) Voltage clamp analysis of membrane currents in larval muscle fibers of *Drosophila*: alteration of potassium currents in Shaker mutants. *J. Neurosci.*, **5**, 2626–2640.
21. Hamill,O.P., Marty,A., Neher,E., Sakmann,B. and Sigworth,F.J. (1981) Improved

patch clamp techniques for high-resolution current recording from cell and cell-free membrane patches. *Pflugers Arch.,* **319**, 85–100.
22. Wu,C.-F., Suzuki,N. and Poo,M.-M. (1983) Dissociated neurons from normal and mutant *Drosophila* larval central nervous system in cell culture. *J. Neurosci.,* **3**, 1888–1899.
23. Sun,Y.-A. and Wu,C.-F. (1985) Genetic alterations of single-channel potassium currents in dissociated CNS neurons of *Drosophila. J. Gen. Physiol.,* **86**, 16a.
24. Byerly,L. (1986) Potassium, sodium and calcium currents in embryonic cultures of *Drosophila* neurons. *Biophys. J.,* **49**, 574a.
25. O'Dowd,D.K. and Aldrich,R.W. (1986) Characterization of sodium channels in cultured *Drosophila* neurons from wild type and temperature sensitive paralytic mutant embryos. *Abstr. Soc. Neurosci.,* **12**, 43.
26. Solc,D.K., Zagotta,W.N. and Aldrich,R.W. (1987) Single-channel and genetic analyses reveal two distinct A-type potassium channels in *Drosophila. Science,* **236**, 1094–1098.
27. Pak,W.L., Grossfield,J. and White,N.V. (1969) Nonphototactic mutants in a study of vision of *Drosophila. Nature,* **222**, 351–254.
28. Hotta,Y. and Benzer,S. (1969) Abnormal electroretinograms in visual mutants of *Drosophila. Nature,* **222**, 354–356.
29. Pak,W.L. (1979) Study of photoreceptor function using *Drosophila* mutants. In *Neurogenetics: Genetic Approaches to the Nervous System.* X.O.Breakfield (ed.), Elsevier North Holland, New York, pp. 67–99.
30. Wu,C.-F. and Pak,W.L. (1975) Quantal basis of photoreceptor spectral sensitivity of *Drosophila melanogaster. J. Gen. Physiol.,* **66**, 149–168.
31. Pirrotta,V. (1986) Cloning *Drosophila* genes. In *Drosophila: A Practical Approach.* D.B.Roberts (ed.), IRL Press, Oxford, pp. 83–110.
32. O'Tousa,J.E., Baehr,W., Martin,R.L., Hirsh,J., Pak,W.L. and Applebury,M.L. (1985) The *Drosophila nina E* gene encodes an opsin. *Cell,* **40**, 839–850.
33. Zuker,C.S., Cowman,A.F. and Rubin,G.M. (1985) Isolation and structure of a rhodopsin gene from *D.melanogaster. Cell,* **40**, 851–858.
34. Salkoff,L., Butler,A., Wei,A., Scavarda,N., Giffen,K., Ifune,C., Goodman,R. and Mandel,G. (1987) Genomic organization and deduced amino acid sequence of a putative sodium channel gene in *Drosophila. Science,* **237**, 744–749.
35. Hermans-Borgemeyer,I., Zopf,D., Ryseck,R.P., Hovemann,B., Betz,H. and Gundelfinger,E.D. (1986) Primary structure of a developmentally regulated nicotinic acetylcholine receptor protein from *Drosophila. EMBO J.,* **5**, 1503–1508.
36. Wadsworth,S.C., Rosenthal,L.S., Kammermeyer,K.L., Potter,M.B. and Nelson,D.J. (1988) Expression of a *Drosophila* acetylcholine receptor-related gene in the central nervous system. *Mol. Cell Biol.,* **8**, 778–785.
37. Itoh,N., Slemmon,J.R., Hawke,D.H., Williamson,R., Morita,E., Itakura,K., Roberts,E., Shively,J.E., Crawford,G.D. and Salvaterra,P.M. (1986) Cloning of *Drosophila* choline acetyltransferase cDNA. *Proc. Natl. Acad. Sci. USA,* **83**, 4081–4085.
38. Zipursky,S.L., Venkatesh,T.R. and Benzer,S. (1985) From monoclonal antibody to gene for a neuron-specific glycoprotein in *Drosophila. Proc. Natl. Acad. Sci. USA,* **82**, 1855–1859.
39. Patel,N.H., Snow,P.M. and Goodman,C.S. (1987) Characterization and cloning of fasciclin III: a glycoprotein expressed on a subset of neurons and axon pathways in *Drosophila. Cell,* **48**, 975–988.
40. Lindsley,D.L. and Grell,E.H.(1968) *Genetic Variations of Drosophila melanogaster,* Carnegie Institution of Washington, Publication no. 627.
41. Carroll,S.B. and Laughon,A. (1987) Production and purification of polyclonal antibodies to the foreign segment of β-galactosidase fusion proteins. In *DNA Cloning, Volume III: A Practical Approach.* D.M.Glover (ed.), IRL Press, Oxford, pp. 89–111.
42. Bender,W., Spierer,P. and Hogness,D.S. (1983) Chromosomal walking and jumping to isolate DNA from the *Ace* and *rosy* loci and the bithorax complex in *Drosophila melanogaster. J. Mol. Biol.,* **168**, 17–33.
43. Kamb,A., Iverson,L.E. and Tanouye,M.A. (1987) Molecular characterization of *Shaker,* a *Drosophila* gene that encodes a potassium channel. *Cell,* **50**, 405–413.
44. Papazian,D.M., Schwarz,T.L., Tempel,B.L., Jan,Y.N. and Jan,L.Y. (1987) Cloning of genomic and complementary DNA from *Shaker,* a putative potassium channel

gene from *Drosophila. Science,* **237**, 749–753.
45. Baumann,A., Krah-Jentgens,I., Muller,R., Muller-Holtkamp,F., Seidel,R., Kecskemethy,N., Casal,J., Ferrus,A. and Pongs,O. (1987) Molecular organization of the maternal effect region of the Shaker complex of *Drosophila*: characterization of an I_A channel transcript with homology to vertebrate Na^+ channel. *EMBO J.,* **6**, 3419–3429.
46. Drysdale,R. and Ganetzky,B. (1987) Molecular analysis of *eag*: a gene affecting potassium channels in *Drosophila. Abs. Soc. Neurosci.,* **13**, 579.
47. Kidwell,M. (1986) P-M mutagenesis. In *Drosophila: A Practical Approach.* D.B.Roberts (ed.), IRL Press, Oxford, pp. 59–81.
48. Ganetzky,B., Loughney,K. and Wu,C.-F. (1986) Analysis of mutations affecting sodium channels in *Drosophila*. In *Tetrodotoxin, Saxitoxin and the Molecular Biology of the Sodium Channel.* C.Y.Kao and S.R.Levinson (eds), New York Academy of Sciences, New York, pp. 325–337.
49. Greenspan,R.J. (1980) Mutations of choline acetyltransferase and associated neural defects in *Drosophila melanogaster. J. Comp. Physiol.,* **137**, 83–92.
50. Hall,J.C. and Kankel,D.R. (1976) Genetics of acetylcholinesterase in *Drosophila melanogaster. Genetics,* **83**, 517–535.
51. Lindsley,D.L., Sandler,L., Baker,B.S., Carpenter,A.T.C., Denell,R.E., Hall,J.C., Jacobs,P.A., Miklos,G.L.G., Davis,B.K., Gethmann,R.C., Hardy,R.W., Hessler,A., Miller,S.M., Nozawa,H., Parry,D.M. and Gould-Somero,M. (1972) Segmental aneuploidy and the genetic gross structure of the *Drosophila* genome. *Genetics,* **71**, 157–184.
52. Ganetzky,B. and Wu,C.-F. (1986) Neurogenetics of membrane excitability in *Drosophila. Annu. Rev. Genet.,* **20**, 13–44.
53. Spradling,A. (1986) P element-mediated transformation. In *Drosophila: A Practical Approach.* D.B.Roberts (ed.), IRL Press, Oxford, pp. 175–197.
54. Spradling,A.G. and Rubin,G.M. (1983) The effect of chromosomal position on the expression of the *Drosophila* xanthine dehydrogenase gene. *Cell,* **34**, 47–57.
55. Scholnick,S.B., Morgan,B.A. and Hirsh,J. (1983) The cloned dopa decarboxylase gene is developmentally regulated when reintegrated into the *Drosophila* genome. *Cell,* **34**, 37–45.
56. Goldberg,D.A., Posakony,J.W. and Maniatis,T. (1983) Correct developmental expression of a cloned alcohol dehydrogenase gene transduced into the *Drosophila* germ line. *Cell,* **34**, 59–73.
57. Scholnick,S.B., Bray,S.J., Morgan,B.A., McCormick,C.A. and Hirsh,J. (1986) CNS and hypoderm regulatory elements of the *Drosophila melanogaster* dopa decarboxylase gene. *Science,* **234**, 998–1002.
58. Florey,E. (1963) Acetylcholine in the invertebrate nervous system. *Can. J. Biochem. Physiol.,* **41**, 2619–2626.
59. Pitman,R.M. (1971) Transmitter substances in insects: a review. *Comp. Gen. Pharmacol.,* **2**, 347–371.
60. Gerschenfeld,H.M. (1973) Chemical transmission in invertebrate central nervous systems and neuromuscular junctions. *Physiol. Rev.,* **53**, 1–119.
61. Sattelle,D.B. (1980) Acetylcholine receptors of insects. *Adv. Insect Physiol.,* **15**, 215–315.
62. Dewhurst,S.A., McCaman,R.E. and Kaplan,W.D. (1970) The time course of development of acetylcholinesterase and choline acetyltransferase in *Drosophila melanogaster. Biochem. Genet.,* **4**, 499–508.
63. Salvaterra,P.M. and McCaman,R.E. (1985) Choline acetyltransferase and acetylcholine levels in *Drosophila melanogaster*. A study using two temperature-sensitive mutations. *J. Neurosci.,* **5**, 903–910.
64. Schmidt-Nielson,B.K., Gepner,J.I., Teng,N.N.H. and Hall,L.M. (1977) Characterization of an alpha-bungarotoxin binding component from *Drosophila melanogaster. J. Neurochem.,* **29**, 1013–1031.
65. Rudloff,E., Himenez,F. and Bartels,F. (1980) Purification and properties of the nicotinic acetylcholine receptors of *Drosophila melanogaster*. In *Receptors for Neurotransmitter, Hormones and Pheromones in Insects.* D.B.Satelle, L.M.Hall and J.G.Hildebrand (eds), Elsevier, Amsterdam, pp. 85–92.
66. Dudai,Y. (1980) Cholinergic receptors of *Drosophila*. In *Receptors for Neurotransmitters, Hormones and Pheromones in Insects.* D.B.Satelle, L.M.Hall and

J.G.Hildebrand (eds), Elsevier, Amsterdam, pp. 93–110.
67. Wu,C.-F., Young,S.H. and Tanouye,M.A. (1983) Single channel recording of alpha-bungarotoxin resistant acetylcholine channels in dissociated CNS neurons of *Drosophila*. *Soc. Neurosci. Abs,* **9**, 507.
68. Greenspan,R.J., Finn,J.A. and Hall,J.C. (1980) Acetylcholinesterase mutants in *Drosophila* and their effects on the structure and function of the nervous system. *J. Comp. Neurol.,* **189**, 741–774.
69. Itoh,N., Slemmon,J.R., Hawke,D.H., Williamson,R., Morita,E., Itakura,K., Roberts,E., Shively,J.E., Crawford,G.D. and Salvaterra,P.M. (1986) Cloning of *Drosophila* choline acetyltransferase cDNA. *Proc. Natl. Acad. Sci. USA,* **83**, 4081–4085.
70. Hall,L.M.C. and Spierer,P. (1986) The *Ace* locus of *Drosophila melanogaster*: structural gene for acetylcholinesterase with an unusual 5' leader. *EMBO J.,* **5**, 2949–2954.
71. Gundelfinger,E.D., Hermans-Borgemeyer,I., Zopf,D., Sawruk,E. and Betz,H. (1986) Characterization of the mRNA and the gene of a putative neuronal nicotinic acetylcholine receptor protein from *Drosophila*. In *Nicotinic Acetylcholine Receptor*. A.Maelicke (ed.), Springer-Verlag, Berlin, p. 437–446.
72. Papazian,D.M., Schwarz,T.L., Tempel,B.L., Timpe,L.C. and Jan,L.Y. (1988) Ion channels in *Drosophila*. *Annu. Rev. Physiol.,* **50**, 379–394.
73. Gorczyca,M. and Hall,J.C. (1984) Identification of a cholinergic synapse in the giant fiber pathway of *Drosophila* using conditional mutations of acetylcholine synthesis. *J. Neurogenet.,* **1**, 289–313.
74. Slemmon,J.R., Salvaterra,P.M., Crawford,G.D. and Roberts,E. (1982) Purification of choline acetyltransferase from *Drosophila melanogaster*. *J. Biol. Chem.,* **257**, 3847–3852.
75. Crawford,G., Slemmon,J.R. and Salvaterra,P.M. (1982) Monoclonal antibodies selective for *Drosophila melanogaster* choline acetyltransferase. *J. Biol. Chem.,* **257**, 3853–3856.
76. Gausz,J., Hall,L.M.C., Spierer,A. and Spierer,P. (1986) Molecular genetics of the *rosy-Ace* region of *Drosophila melanogaster*. *Genetics,* **112**, 65–78.
77. MacPhee-Quigley,K., Taylor,P. and Taylor,S.S. (1985) Primary structures of the catalytic subunits from two molecular forms of acetylcholinesterase. *J. Biol. Chem.,* **260**, 12185–12189.
78. Boulter,J., Lutyen,W., Evans,K., Mason,B., Ballivet,M., Goldman,D., Stengelin,S., Martin,G., Heinemann,S. and Patrick,J. (1986) Isolation of a cDNA clone coding for a possible neural nicotinic acetylcholine receptor alpha-subunit. *Nature,* **319**, 368–374.
79. Stryer,L. (1987) The molecules of visual excitation. *Sci. Am.,* **257**, 42–50.
80. Heisenberg,M. and Wolf,R. (1984) *Vision in Drosophila: Genetics of Microbehavior*. Springer-Verlag, Berlin.
81. Scavarda,N.J., O'Tousa,J. and Pak,W.L. (1983) *Drosophila* locus with gene dosage effects on rhodopsin. *Proc. Natl. Acad. Sci. USA,* **80**, 4441–4445.
82. Nathans,J. and Hogness,D.S. (1983) Isolation sequence analysis and intron–exon arrangement of the gene encoding bovine rhodopsin. *Cell,* **34**, 807–814.
83. Cowman,A.F., Zuker,C.S. and Rubin,G.M. (1986) An opsin gene expressed in only one photoreceptor cell type of the *Drosophila* eye. *Cell,* **44**, 705–710.
84. Zuker,C.S., Montell,C., Jones,K., Laverty,T. and Rubin,G.M. (1987) A rhodopsin gene expressed in photoreceptor cell R7 of the *Drosophila* eye: homologies with other signal transducing molecules. *J. Neurosci.,* **7**, 1550–1557.
85. Fryxell,K.J. and Meyerowitz,E.M. (1987) An opsin gene that is expressed only in the R7 photoreceptor cell of *Drosophila*. *EMBO J.,* **6**, 443–451.
86. Montell,C., Jones,K., Zuker,C.S. and Rubin,G.M. (1987) A second opsin gene expressed in the ultraviolet sensitive R7 photoreceptor cells of *Drosophila melanogaster*. *J. Neurosci.,* **7**, 1558–1566.
87. Cosens,D. and Manning,A. (1969) Abnormal electroretinogram from a *Drosophila* mutant. *Nature,* **224**, 285–287.
88. Minke,B., Wu,C.-F. and Pak,W.L. (1975) Induction of photoreceptor voltage noise in the dark in *Drosophila* mutant. *Nature,* **258**, 84–87.
89. Montell,C., Jones,K., Hafen,E. and Rubin,G.M. (1985) Rescue of the *Drosophila*

phototransduction mutation *trp* by germline transformation. *Science,* **230**, 1040–1043.
90. Wong,F., Hokanson,M. and Chang,L.T. (1985) Molecular basis of an inherited retinal defect in *Drosophila. Invest. Opthalmol. Vis. Sci.,* **26**, 243–246.
91. Pak,W.L., Grossfield,J. and Arnold,K. (1970) Mutants of the visual pathway of *Drosophila melanogaster. Nature,* **227**, 518–520.
92. Hotta,Y. and Benzer,S. (1970) Genetic dissection of the *Drosophila* nervous system by means of mosaics. *Proc. Natl. Acad. Sci. USA,* **67**, 1156–1163.
93. Ostroy,S.E., Wilson,M. and Pak,W.L. (1974) *Drosophila* rhodopsin: photochemistry, extraction and differences in the *norpA*^{P12} phototransduction mutant. *Biochem. Biophys. Res. Commun.,* **59**, 960–966.
94. Pak,W.L. and Lidington,K.J. (1974) Fast electrical potential from a long-lived wavelength photoproduct of fly visual pigment. *J. Gen. Physiol.,* **63**, 740–756.
95. Inoue,H., Yoshioka,T. and Hotta,Y. (1985) A genetic study of inositol triphosphate involvement in phototransduction using *Drosophila* mutants. *Biochem. Biophys. Res. Commun.,* **132**, 513–519.
96. Fein,A., Payne,R., Corson,D.W., Berridge,M.J. and Irvine,R.F. (1984) Photoreceptor excitation and adaptation by inositol 1,4,5-triphosphate. *Nature,* **311**, 157–160.
97. Brown,J.E., Rubin,L.J., Ghalayini,A.J., Tarver,A.P., Irvine,R.F., Berridge,M.J. and Anderson,R.E. (1984) *myo*-inositol polyphosphate may be a messenger for visual excitation in *Limulus* photoreceptors. *Nature,* **311**, 160–163.
98. Hille,B. (1985) *Ionic Channels of Excitable Membranes*. Sinauer, Sunderland.
99. Salkoff,L., Butler,A., Scavarda,N. and Wei,A. (1987) Nucleotide sequence of the putative sodium channel gene from *Drosophila*: the four homologous domains. *Nucleic Acids Res.,* **15**, 8569–8572.
100. Noda,M., Shimizu,S., Tanabe,T., Takai,T., Kayano,T., Ikeda,T., Takahashi,H., Nakayama,H., Kanaoka,Y., Minamino,N., Kangawa,K., Matsuo,H., Raftery,M.A., Hirose,T., Inayama,S., Hayashida,H., Miyata,T. and Numa,S. (1984) Primary structure of *Electrophorus electricus* sodium channel deduced from cDNA sequence. *Nature,* **312**, 121–127.
101. Noda,M., Ikeda,T., Kayano,T., Suzuki,H., Takeshima,H., Kurasaki,M., Takahashi,H. and Numa,S. (1986) Existence of distinct sodium channel messenger RNAs in rat brain. *Nature,* **320**, 188–192.
102. Jackson,F.R., Wilson,S.D., Strichartz,G.R. and Hall,L.M. (1984) Two types of mutants affecting voltage-sensitive sodium channels in *Drosophila melanogaster. Nature,* **308**, 189–191.
103. Jackson,F.R., Gitschier,J., Strichartz,G.R. and Hall,L.M. (1985) Genetic modifications of voltage-sensitive sodium channels in *Drosophila*: gene dosage studies of the seizure locus. *J. Neurosci.,* **5**, 1144–1151.
104. Ganetzky,B. and Wu,C.-F. (1982) Indirect suppression involving behavioral mutants with altered nerve excitability in *Drosophila melanogaster. Genetics,* **100**, 597–614.
105. Suzuki,N. and Wu,C.-F. (1984) Altered sensitivity to sodium channel-specific neurotoxins in cultured neurons from temperature-sensitive paralytic mutants of *Drosophila. J. Neurogenet.,* **1**, 225–238.
106. Kauvar,L.M. (1982) Reduced [^{3}H]-Tetrodotoxin binding in the *nap*^{ts} paralytic mutant of *Drosophila. Mol. Gen. Genet.,* **187**, 172–173.
107. West,G.J. and Catterall,W.A. (1979) Selection of variant neuroblastoma clones with missing or altered sodium channels. *Proc. Natl. Acad. Sci. USA,* **76**, 4136–4140.
108. Suzuki,D.T., Grifliatti,T. and Williamson,R. (1971) Temperature-sensitive mutants in *Drosophila melanogaster,* VII. A mutation (*para*^{ts}) causing reversible adult paralysis. *Proc. Natl. Acad. Sci. USA,* **68**, 890–893.
109. O'Dowd,D.K., Germeraad,S. and Aldrich,R.W. (1987) Expression of sodium currents in embryonic *Drosophila* neurons: differential reduction by alleles of the *para* locus. *Abs Soc. Neurosci.,* **13**, 577.
110. Kulkarni,S.J. and Padhye,A. (1982) Temperature-sensitive paralytic mutations on the second and third chromosomes of *Drosophila melanogaster. Genet. Res.,* **40**, 191–199.
111. Jackson,F.R., Wilson,S.D. and Hall,L.M. (1986) The *tip-E* mutation of *Drosophila* decreases saxitoxin binding and interacts with other mutations affecting nerve membrane excitability. *J. Neurogenet.,* **3**, 1–17.

112. Kasbekar,D., Nelson,J. and Hall,L.M. (1987) Enhancer of seizure: a new genetic locus in *Drosophila melanogaster* defined by interactions with temperature-sensitive paralytic mutations. *Genetics*, **116**, 423–431.
113. Elkins,T.T. (1986) *Electrophysiological Analyses of Drosophila Behavioral Mutants.* PhD thesis. The University of Wisconsin, Madison, WI.
114. Ganetzky,B. (1984) Genetic studies of membrane excitability in *Drosophila*: lethal interaction between two temperature-sensitive paralytic mutations. *Genetics*, **108**, 897–911.
115. Burg,M.G. and Wu,C.-F. (1986) Differentiation and central projections of peripheral sensory cells with action-potential block in *Drosophila* mosaics. *J. Neurosci.*, **6**, 2968–2976.
116. Adams,D.J., Stephen,J.S. and Thompson,S.H. (1980) Ionic currents in Molluscan soma. *Annu. Rev. Neurosci.*, **3**, 141–167.
117. Salkoff,L. (1983) *Drosophila* mutants reveal two components of fast outward currents. *Nature*, **302**, 249–251.
118. Wei,A. and Salkoff,L. (1986) Occult *Drosophila* calcium channels and twinning of calcium and voltage-activated potassium channels. *Science*, **233**, 780–782.
119. Gho,M. and Mallart,A. (1986) Two distinct calcium-activated potassium currents in larval muscle fibres of *Drosophila melanogaster*. *Pflugers Arch.*, **407**, 526–533.
120. Singh,S. and Wu,C.-F. (1987) Genetic and pharmacological separation of four potassium currents in *Drosophila* larvae. *Abs Soc. Neurosci.*, **13**, 579.
121. Stevens,C.F. (1987) Channel families in the brain. *Nature*, **328**, 198–199.
122. Kaplan,W.D. and Trout,W.E. III. (1969) The behavior of four neurological mutants of *Drosophila*. *Genetics*, **61**, 399–409.
123. Jan,Y.N., Jan,L.Y. and Dennis,M.J. (1977) Two mutations of synaptic transmission in *Drosophila*. *Proc. R. Soc. Lond. B*, **198**, 87–108.
124. Salkoff,L. (1983) Genetic and voltage clamp analysis of a *Drosophila* potassium channel. *Cold Spring Harbor Symp. Quant. Biol.*, **48**, 221–231.
125. Haugland,F.N. (1987) *A Voltage-clamp Analysis of Membrane Potassium Currents in Larval Muscle Fibers of the Shaker Mutants of Drosophila.* PhD thesis. The University of Iowa, Iowa City, IA.
126. Salkoff,L.B. and Tanouye,M.A. (1986) Genetics of ion channels. *Physiol. Rev.*, **66**, 302–329.
127. Tanouye,M.A., Kamb,C.A., Iverson,L.E. and Salkoff,L. (1986) Genetics and molecular biology of ionic channels in *Drosophila*. *Annu. Rev. Neurosci.*, **9**, 255–276.
128. Tempel,B.L., Papazian,D.M., Schwarz,T.L., Jan,Y.N. and Jan,L.Y. (1987) Sequence of a probable potassium channel component encoded at Shaker locus of *Drosophila*. *Science*, **237**, 770–775.
129. Tanabe,T., Takeshima,H., Mikami,A., Flockerzi,V., Takahashi,H., Kangawa,K., Kojima,M., Matsuo,H., Hirose,T. and Numa,S. (1987) Primary structure of the receptor for calcium channel blockers from skeletal muscle. *Nature*, **328**, 313–318.
130. Ganetzky,B. and Wu,C.-F. (1983) Neurogenetic analysis of potassium currents in *Drosophila*: synergistic effects on neuromuscular transmission in double mutants. *J. Neurogenet.*, **1**, 17–28.
131. Timpe,L.C. and Jan,L.Y. (1987) Gene dosage and complementation analysis of the Shaker locus in *Drosophila*. *J. Neurosci.*, **7**, 1307–1317.
132. Ganetzky,B. and Wu,C.-F. (1985) Genes and membrane excitability in *Drosophila*. *Trends Neurosci.*, **8**, 322–326.
133. Drysdale,R. and Ganetzky,B. (1985) Cloning of a gene affecting potassium channels in *Drosophila*. *Soc. Neurosci. Abs*, **11**, 788.
134. Stern,M. and Ganetzky,B. (1989) Altered synaptic transmission in *Drosophila Hyperkinetic* mutants. *J. Neurogenet.*, in press.
135. Elkins,T. and Ganetzky,B. (1988) The roles of potassium currents in *Drosophila* flight muscles. *J. Neurosci.*, **8**, 428–434.
136. Schaefer,W.H., Hinrichsen,R.D., Burgess-Cassler,A., Kung,C., Blair,I.A. and Watterson,D.M. (1987) A mutant *Paramecium* with a defective calcium-dependent potassium conductance has an altered calmodulin: a nonlethal selective alteration in calmodulin regulation. *Proc. Natl. Acad. Sci. USA*, **84**, 3931–3935.
137. Mishina,M., Tobimatsu,T., Imoto,K., Tanaka,K., Fujita,Y., Fukuda,K., Kurasaki,M., Takahashi,H., Morimoto,Y., Hirose,T., Inayama,S., Takahashi,T., Kuno,M. and

Numa,S. (1985) Location of functional regions of acetylcholine receptor α-subunit by site directed mutagenesis. *Nature,* **313**, 364–369.
138. Cosens,D.J. and Perry,M.M. (1972) The fine structure of the eye of a visual mutant, A-type of *Drosophila melanogaster. J. Insect Physiol.,* **18**, 1773–1786.
139. Catterall,W.A. (1986) Voltage-dependent gating of sodium channels: correlating structure and function. *Trends Neurosci.,* **9**, 7–10.

3

Molecular genetics of acetylcholine receptor-channels
Toni Claudio

1. Introduction

The application of molecular genetics techniques to the study of ion channels began with studies on the nicotinic acetylcholine receptor (AChR). The long-standing interest in the function of this molecule has resulted in an intense and continued study of its properties. Two factors have contributed significantly to the successful isolation and characterization of the AChR: a source of material from which large quantities of AChR could be isolated, and very specific ligands for monitoring and isolating purified AChR. Information about this protein has been steadily accumulating for almost 20 years and many excellent reviews are available that deal with the biochemistry (1–7), pharmacology (8–11), electrophysiology (12–15), biosynthesis (16–18), developmental regulation (12–23), immunology (24–26), structure (27–29) and, most recently, the molecular biology of the AChR (30,31). A tremendous amount of information has been amassed since crude preparations of AChR-containing mRNA were first microinjected into *Xenopus laevis* oocytes in 1981 (32). The cloning of cDNAs that encode the subunits of this receptor led to the deduction of their amino acid sequences. This in turn has led to a wide range of experimentation to test structural and functional predictions about the molecule. This review will focus upon the AChR and will extensively discuss information that has been gained concerning its structure and function through the application of recombinant DNA and molecular genetic approaches. This work exemplifies the approaches that will undoubtedly be applied to other ligand-gated receptor-channels (the glycine and the γ-aminobutyric acid receptors) as well as voltage-gated channels (sodium, calcium and potassium) which have recently been cloned.

The amino acid sequences deduced from these more recently cloned channels indicate that channels with similar functions have similar structures. Three ligand-gated channels have several structural features in common as do three voltage-gated channels, representing two superfamilies of channels. The 'common function – common structure' concept is certainly not a novel observation, nor is the idea that the members of a family of proteins are probably derived from a single ancestral gene. Such commonalities among proteins and evolutionary relationships have been widely reported, especially for a large number of enzymes (33). The belief is that it is easier to duplicate and then modify genes in order to produce divergence in proteins rather than to reconstruct a new protein *de novo*.

I will first give a brief introduction to these two superfamilies of channels and then review in detail the extensive studies that have been carried out on the AChR.

1.1 Ligand-gated receptor-channels

Traditionally, the ligand-gated channels have been classified according to the ligands which bind to them. The mechanism of channel-activation for this group of receptor-channels is similar. Upon binding by the appropriate neurotransmitter or agonist, channels open and ions flow through the pore. Thus far, three ligand-gated receptor-channels have been cloned: the acetylcholine, the glycine (34) and the γ-aminobutyric acid ($GABA_A$) receptors (35).

1.1.1 Acetylcholine receptors

The AChR is a complex multisubunit protein that is inherently involved in synaptic transmission, synaptogenesis, and the autoimmune disease, myasthenia gravis. Because of its variety of roles, this molecule has been of interest to a wide variety of investigators including protein chemists, neurobiologists and immunobiologists. The AChR is a ligand-gated ionic channel located in the post-synaptic membrane of the vertebrate neuromuscular junction that mediates synaptic transmission between nerve and muscle. In response to nerve stimulation, vesicles containing the neurotransmitter acetylcholine (ACh) fuse to the plasma membrane of the nerve terminal (pre-synaptic membrane) and ACh is released into the synaptic cleft (36). The quantum release of ACh per vesicle is about 10^4 molecules. ACh molecules diffuse about 50 nm to the post-synaptic membrane where they bind to AChRs. The AChR upon binding ACh undergoes a conformational change which results in the opening of a cation-selective channel through which approximately 10^7 ions flow per second. The simultaneous release of about 200 quanta results in the depolarization of the muscle cell which eventually leads to muscle contraction. The morphology of the post-synaptic membrane is quite remarkable and highly specialized (37). Invaginations of the plasma membrane form the junctional folds which are 0.5–1 μm deep. The 'junctional' AChRs are expressed on the tops of these folds at a site density of about 10^4 receptors per μm^2, whereas acetylcholinesterase (responsible for hydrolyzing ACh) extends from the top to the bottom of the fold and has a site density of about 2×10^3 molecules per μm^2.

The properties of the AChR will be described in detail in Section 2. Briefly, it is an intrinsic membrane glycoprotein composed of four different subunits in a pentameric complex, $\alpha_2\beta\gamma\delta$, with an overall molecular mass of approximately 250 kd. Each subunit spans the membrane a number of times and several models have been proposed to account for the distribution of different domains of the protein within the membrane. The alternative models are discussed extensively in Section 4.5. cDNAs for the AChR subunits were first cloned from mRNA isolated from the electric organ of *Torpedo* (Section 3). These clones were used in turn to isolate cDNAs corresponding to both muscle-like nicotinic AChRs (Sections 5, 6 and 9) and neuronal AChRs (Section 10) from several species, making this the best characterized family of ligand-gated receptor-channels.

1.1.2 GABA and glycine receptors

Both GABA and glycine are inhibitory neurotransmitters, the binding of which result in the opening of integral membrane chloride channels. The influx of anions leads to hyperpolarization of the post-synaptic membrane and inhibition of neuronal firing. GABA and glycine are the major inhibitory neurotransmitters in the central nervous system (CNS) of vertebrates and many invertebrates. In mammals, GABAergic synapses are abundant in the brain, whereas glycinergic synapses are abundant in the brain stem and spinal cord. The convulsive alkaloid, strychnine, is the most potent specific antagonist known for glycine receptors (reviewed in ref. 38) and it has been used extensively for analyzing and purifying this receptor. Two pharmacologically and functionally distinct classes of GABA receptors (reviewed in ref. 39) have been defined, $GABA_A$ and $GABA_B$. The $GABA_A$ receptor is linked directly to a chloride channel whereas the $GABA_B$ receptor is linked via a G protein and/or cAMP to potassium and calcium channels. It is the 'classical' GABA receptor ($GABA_A$) which has been cloned (35) and which will be discussed in Section 11.1.3. Several pharmacologically significant drugs interact with this receptor-channel, such as anxiolytics (benzodiazepines), anticonvulsants (barbiturates) and convulsants (picrotoxin). Both $GABA_A$ and glycine receptors are heterologous multisubunit complexes; however, the subunit composition of these channels is not nearly as well-defined as that of the AChR. Like the AChR, the molecular masses of the GABA and glycine receptors are in the order of 250 kd. As will be discussed in Sections 11.1.2 and 11.1.3, some very interesting findings concerning the structure and function of these channels are emerging now that subunit cDNAs have been isolated and expressed.

1.2 Voltage-gated channels

In contrast to the ligand-gated channels, the voltage-gated channels traditionally have been classified according to the major permeant ion which flows through the channel and by properties related to the manner in which the channels are activated or inactivated. The mechanism of channel activation is also quite distinct between ligand-gated and voltage-gated channels. Voltage-gated channels open in response to a change in the potential across the membrane and they are characterized by high ion selectivity. The voltage-sensitive ion channels play a fundamental role in the generation and propagation of action potentials in electrically excitable cells. They perform important roles in signal transduction in other cell types, as well (40,41). Several voltage-gated channels have recently been cloned and sequenced. These include: sodium channels from *Electrophorus electricus* (42), rat brain (43), *Drosophila* (44,45) and rat skeletal muscle (46); a dihydropyridine-sensitive calcium channel from rabbit skeletal muscle (47,48); and potassium channels from *Drosophila* (49–55), mouse brain (56) and rat brain (57).

1.2.1 Sodium channels

For recent reviews on Na^+ channels, see refs 58–60. Voltage-sensitive Na^+ channels have been purified from diverse tissue sources such as the electric organ from eel, cardiac muscle and brain. Depending on the tissue from which it was isolated, the channel appears to be composed of from one to four subunits. Regardless of the tissue source, these channels are composed of at least one large polypeptide with a molecular mass of approximately 250 kd. The cloning of several Na^+ channels is discussed in Section 11.2.1.

1.2.2 Calcium channels

The calcium channels are a heterogeneous group of molecules which historically have been grouped into voltage-sensitive and receptor-operated channels (61–63). These channels have been found in muscle, neuronal, secretory and fibroblast cells. Several types of Ca^{2+} channels have been distinguished by their pharmacological and physiological properties. The voltage-dependent Ca^{2+} channels alone can be further divided into at least three categories, often referred to as the L- (long-lasting), T- (transient) and N- (neither) types. In the L-type channel, ion flux can be modulated by the Ca^{2+} antagonist, 1,4-dihydropyridine (DHP). The composition of the DHP-sensitive Ca^{2+} channel is unclear. It may be composed of five different subunits, containing at least two large polypeptide subunits, α_1 (155–170 kd) and α_2 (135–150 kd). The α_1 subunit binds Ca^{2+} antagonists, including DHP. Complementary DNAs for the α_1 (47,48) and α_2 (48) polypeptides were recently cloned from rabbit skeletal muscle. Sequence analysis of the α_1 polypeptide revealed that it had striking similarities with the large polypeptides of Na^+ channels (see Section 11.2.2) whereas sequence analysis of the α_2 polypeptide revealed no homologies to other known proteins.

1.2.3 Potassium channels

The potassium-selective channels appear to be the most heterogeneous group of ion channels characterized thus far (40,64). Diversity in channels is displayed between different tissues as well as within the same tissue. Some channels open in response to membrane depolarization, others to membrane hyperpolarization, and some to an increase in intracellular Ca^{2+}. Some K^+ channels are known to respond to transmitter binding or second messengers and show differences in their kinetic properties and single-channel conductances. The lack of a tissue source rich in a single population of channels plus the lack of high-affinity, high-specificity ligands and antibodies has hindered the isolation and characterization of these important channels. The first K^+ channel cDNA clones were isolated from *Drosophila* using a genetic approach (65), and these in turn were used to isolate homologous clones from mouse and rat brain (see Section 11.2.3).

2. Background to the AChR

I will now review studies of the AChR in some detail, since they exemplify the approaches that can be followed to facilitate the characterization of an ion channel once cloned DNA sequences are available. First of all, however, it is valuable to review the properties of this receptor as revealed by more traditional approaches.

The best source of AChRs has been from electric fish. The electric organs of the elasmobranch *Torpedo* (a marine ray) and the teleost *Electrophorus* (a freshwater eel) are both composed of cells called electrocytes or electroplaques which are modified muscle cells that have lost their ability to contract. The electrocytes from *Torpedo* and *Electrophorus* are, however, quite distinct (66). *Torpedo* electrocytes have a diameter of 5–7 mm and a thickness of only 7–20 μm. They are arranged in columns of about 1000 cells. The ventral surfaces are highly and multiply innervated, and the cells are electrically inexcitable. The transcellular potential across each electroplaque in *Torpedo* is due solely to the endplate depolarization, with the result that a discharge from this electric organ is only about 60 V. In contrast, *Electrophorus* electrocytes are approximately box-shaped, of dimensions 10 × 2 × 0.1 mm. Their caudal surfaces are innervated, and the cells are electrically excitable. Several thousand electroplaques in series constitute a row in which the voltages of about 150 mV produced across individual cells by action potentials adds up to approximately 600 V. The unique anatomical properties of these two electric tissues have made *Electrophorus* the best source for isolating intact cells and for the isolation of sodium channels. In contrast, *Torpedo* has been the best source of AChRs. Two 1 kg electric organs can be obtained from a large *Torpedo*, from which almost 0.2 g of pure receptor can be isolated. Needless to say, most of the biochemistry, pharmacology and structural work has been conducted on *Torpedo* AChR.

2.1 Composition and structure

The AChR is an intrinsic membrane glycoprotein composed of four different subunits α, β, γ and δ, with apparent molecular weights of approximately 40, 50, 60 and 65 kd, respectively. The functional molecule is a pentameric complex of $\alpha_2\beta\gamma\delta$ subunits which has a molecular weight of approximately 250 kd and migrates on sucrose gradients with a sedimentation coefficient of 9S. ACh binds to the α subunits of the receptor and thus there are two ACh binding sites per AChR molecule. A group of curaremimetic neurotoxins isolated from the venom of snakes belonging to the families *Elapidae* (cobras, kraits, mambas and coral snakes) and *Hyrophidae* (sea snakes) have been shown to bind with high specificity and affinity to nicotinic AChRs. These 7–8 kd toxins are all basic, having a pI of about 9, and block receptor function. The elapid snake neurotoxin, α-bungarotoxin (BuTx) (67), has been widely used in the isolation, purification and characterization of the AChR. The binding of this molecule is essentially irreversible having a K_D of approximately 10^{-11} M. It can be radiolabeled without loss of binding function. With *Torpedo* AChRs, but not AChRs isolated from other sources, pairs of pentameric complexes, often referred to as 9S 'monomers', are covalently linked through cysteine disulfide bonds between δ subunits to yield 500 kd AChR 'dimers' (13S). Thus far, no differences in functional properties between monomers and dimers have been detected. Some of the post-translational modifications of the AChR that are known to occur include glycosylation, fatty acylation, phosphorylation and disulfide bond formation (1–7).

The structural features of the AChR have been determined using electron images of stained molecules (reviewed in refs 68,69,70), neutron scattering (71), X-ray diffraction (69), and metal replicas of freeze-fractured or freeze-etched AChRs (70,72). The subunits of the AChR form a rosette-like structure with a central pore. Three-dimensional electron image analysis of tubular crystals of AChR at a resolution of 2.5 nm has indicated that the subunits are arranged regularly around a central axis in near pentagonal symmetry (73). The molecule is approximately 12 nm in length with about 5.5 nm extending on the extracellular surface and about 2 nm extending on the cytoplasmic surface. The outside diameter of the molecule is about 8 nm. A central water-filled pore extends the length of the molecule and has a diameter on the extracellular surface of about 2.5 nm but is constricted to approximately 0.64 nm at or within the lipid bilayer (74).

2.2 Pharmacology

There are three major classes of ligands that interact with the AChR: agonists, competitive antagonists and non-competitive antagonists (inhibitors). Agonists such as ACh, nicotine and carbamylcholine bind to the ACh site and open the channel. Competitive antagonists (of which there are two types) are thought to compete for binding at the ACh site and prevent the opening of the channel. One class of competitive antagonist is readily reversible (dimethyl-*d*-tubocurarine, hexamethonium) while the other is slowly reversible (the curaremimetic toxins). Agonists and competitive antagonists bind to the α subunit of the AChR. Non-competitive antagonists or inhibitors block the receptor response to agonists. A large number of compounds acting in a variety of manners are included in the non-competitive antagonist category. These compounds are primarily organic amines and include certain local anesthetics (lidocaine, trimethisoquin), alkaloid toxins (histrionicotoxin), neuroleptics (chlorpromazine), hallucinogens (phencyclidine), antimalarials (quinacrine), detergents and alcohols. They may act in a number of ways including binding to the open channel and sterically blocking its closure while preventing the passage of ions, or by stabilizing the resting (closed) state, or by promoting desensitization. Binding sites have been localized by affinity labeling which, depending on the non-competitive inhibitor, the source of the receptor and the activation state, have been primarily on α (75), δ (76–78), α and β (79), β and δ (80), or all four subunits (81) (reviewed in ref. 82).

There appear to be at least four general functional states of the receptor: resting (channel closed but activatable), open (active), fast-onset desensitized (occurring in milliseconds) and slow-onset desensitized (occurring in seconds to minutes). In the resting state, the affinity for agonists is low and binding leads to channel opening. With prolonged exposure to agonists the receptor enters the desensitized state where the channel is non-conducting and the affinity for agonists is high. The mechanisms of agonist and antagonist binding, the coupling of binding to the opening and closing of the channel, and the transitions between states are complex processes which are only partially understood (8–11, 13, 15, 83–85).

2.3 Physiology

The most extensive characterization of AChRs has been at the electrophysiological level where AChRs from a wide variety of sources have been investigated. These sources include muscle cell lines, mammalian, avian, amphibian and invertebrate adult and embryonic primary muscle cultures, *Electrophorus* electroctyes, and *Torpedo* AChRs reconstituted into lipid vesicles and planar lipid bilayers. In *Torpedo*, where the most thorough biochemical and structural work has been performed, kinetic and physiological measurements have been obtained from reconstituted AChRs, but it has been impossible to obtain such data from intact electrocytes. In order to compare the structural and biochemical data with the kinetic and physiological data it is imperative that all of the functional properties of AChRs be properly reconstituted in these artificial systems. Functionality is most convincingly tested physiologically, and thus electrophysiology has not only been instrumental in the early studies of AChR function, but it has continued to play a vital role as the definitive assay for channel properties, especially in the employment of molecular genetic approaches to the study of AChRs. Physiological analysis is often called upon to identify novel subunits, to characterize previously uncharacterized receptors, and to interpret the consequences of introducing mutations into genes or cDNAs.

The 'classical' electrophysiological studies involve recording the membrane potential or current responses due to the activity of the population of AChRs at an endplate. From these studies, a number of features of the AChR have been determined. Na^+ and K^+ (and perhaps Ca^{2+}) are the physiologically important permeant ions. However, the channel is permeable to monovalent and divalent cations and even to some small positively charged organic compounds. The most permeant ion is thallium followed by $Cs^+ > Rb^+ > K^+ > Na^+ > Li$. The selectivity among these monovalents is weak as is the selectivity among divalent cations for which the sequence is $Mg^{2+} > Ca^{2+} > Ba^{2+} > Sr^{2+}$ (86). Many other features of the channel and its synaptic environment have also been determined with electrophysiological measurements, namely that

(i) there are approximately 10^4 ACh molecules per quantum,
(ii) there are about 200 quanta released per action potential,
(iii) two ACh molecules are required to bind per receptor for efficient channel opening,
(iv) the rate of ion passage through the channel is about 10^7 ions per second, and
(v) channel opening is much faster than channel closing.

In addition, much has been unraveled concerning the transitions between closed, open and desensitized channels (12,15).

Analyses up to this point were based on the behavior of populations of channels. In 1976, Neher and Sakmann reported the first recordings of currents through single channels (87) and by 1980 they had developed the 'patch-clamp' method by which very tight seals between the glass micropipette and the cell membrane could be attained (88). These are known as 'gigaseals' and have resistances of greater than $10^9\ \Omega$. This allows the study of both cell-attached and cell-free patches of membrane with areas of about 10 μm^2 (89,90). The combined ability to analyze single-channel behavior and to manipulate the external and internal environments of cell-free patches has resulted in major breakthroughs in the ability to characterize channels and receptors. Not only have new channels been identified but the behavior of most channels has been determined to be much more complex than previously thought. It has proved possible using single-channel analysis to determine the time for which the AChR channels remain opened and closed. This has led to detailed kinetic models of the channel's conformational changes. Furthermore, single-channel recordings have enabled the different actions of various agonists to be distinguished, different populations of AChRs to be clearly defined, and the effects of the concentration of agonists on gating (the channel's open and closed times) to be determined (12–15).

2.4 Development

The AChR also displays different properties depending on its location relative to the nerve terminal and depending on developmental time. In the embryonic state, before innervation, AChRs are located diffusely over the surface of the muscle at a surface density of about 20 per μm^2. Upon innervation, the AChRs become tightly clustered underneath nerve terminals where they pack at a density of about 20 000 per μm^2. Such clustering appears to be induced only by cholinergic neurons. In-vitro studies have shown that cholinergic motor-neurons from spinal cord and parasympathetic neurons can induce clusters whereas non-motor-neuron spinal cord cells and sensory neurons cannot. It is also possible to induce clusters with conditioned medium from neuroblastoma–glioma hybrid cells (NG 108-15), rat pheochromocytoma cells (PC12), extracts from spinal cord, extracellular matrix, and a 43 kd glycoprotein recently identified from brain (16,19–23,91–93). A number of mechanisms have been implicated in the process of mobilizing receptors to the synaptic site and establishing clusters. These include (i) passive diffusion of receptors already expressed on the surface of the cell (94; reviewed in ref. 95), (ii) electromigration of receptors to the synaptic site induced by the currents established between AChRs and the nerve terminal at newly forming synapses (95), (iii) localized synthesis and insertion of receptors at the site due to the accumulation of more nuclei at the nerve terminal (96) and/or the increased synthesis of AChR mRNAs from nuclei located at synapses compared with nuclei located some distance from the synapse (97). A variety of compounds have been identified that might be intimately or peripherally involved in establishing, maintaining or organizing receptor clusters. These include neurotrophic factors, basal lamina [the components agrin-150 kd and agrin-95 kd (98) have recently been purified], cytoskeletal elements [β-spectrin (99), α-actinin, vinculin (100,101), filamin (101), talin (102); reviewed in ref. 22], a 43 kd protein (reviewed in ref.

103; see also 100,101,104,105), and metabolic stabilization (19–21,93, 95,103).

Interestingly, muscle activity does not appear to have much regulatory influence on junctional AChRs. In contrast, it has profound effects on extrajunctional receptors. If muscle activity is inhibited by blocking nerve activity (e.g. preventing transmitter release or denervation), blocking action potentials (e.g. with tetrodotoxin, a Na^+-channel blocker), or blocking neuromuscular transmission (e.g. with BuTx or curare), both the rate of AChR synthesis and the number of surface receptors is increased by 50- to 100-fold, a phenomenon that has been termed 'denervation supersensitivity'. These newly expressed receptors appear evenly distributed along the muscle fiber. If the muscle is directly stimulated during such blocking experiments, the appearance of extrajunctional receptors is prevented. If the muscle is electrically stimulated after the appearance of extrajunctional receptors, extrajunctional receptor expression decreases. A few functional and structural differences between junctional and extrajunctional receptors have also been noted. Junctional or adult AChRs turn over more slowly than extrajunctional or embryonic AChRs. In chick, the half-time for turnover of junctional AChRs is about 5 days whereas it is about 30 h for extrajunctional AChRs. Junctional receptors might be slightly more acidic (106), and differences in glycosylation (107) and phosphorylation (108) have been reported. Immunological differences have also been reported. Sera from patients with the autoimmune disease myasthenia gravis can partially inhibit BuTx binding to AChRs from denervated rat muscle (extrajunctional) but not from innervated muscle (junctional) (19), and a class of antibodies present in sera from several myasthenics was found to recognize an epitope that was expressed uniquely on extrajunctional receptors (110). Larger currents are conducted through junctional receptors but the channel open time is shorter. Rat adult endplate AChRs have a single channel conductance of about 50 picosiemens (pS) and a mean channel open time of about 1 ms. These are referred to as fast channels. Rat embryonic muscle AChRs, on the other hand, have a single channel conductance of about 35 pS and a mean channel open time of about 6 ms. These are known as slow channels. Because changes in channel properties appear to occur both with AChRs located at the endplate and in extra-synaptic regions during endplate maturation (111–114; reviewed in ref. 23), the more descriptive and appropriate terminology for junctional and extrajunctional AChRs is probably fast and slow. As will be discussed later (Sections 6 and 8.2.3), a novel AChR subunit has been isolated whose identity and probable function were both determined from the type of channel it formed (fast or slow).

2.5 Immunology

The AChR has been shown to be intimately involved in the autoimmune disease myasthenia gravis (MG). This is a disease occurring with a frequency of 1 in 20 000 and is characterized by skeletal muscle weakness and fatigue. The weakness is due to defective neuromuscular transmission which is now known to be caused by a reduced number of functional AChRs at neuromuscular junctions. The reduction is brought about by antibodies directed against AChRs. Different muscles of the body may be selectively affected. An ocular form of the disease exists resulting in drooping eyelids or double vision; the cranial musculature may be affected resulting in altered facial expressions, impairment in speech, impairments in the ability to chew or swallow; or generalized weakness may occur. It can become life-threatening when the muscles involved in respiration or swallowing are affected. In addition to the adult form, MG also occurs in infancy and childhood (juvenile myasthenia, neonatal myasthenia, congenital myasthenia and familial infantile myasthenia). Although neither the cause nor the cure for this autoimmune disease are known, a great deal of progress has been made delineating the pathological mechanisms of the disease and this knowledge is being used in both diagnosis and therapy. Various therapies have beneficial effects for some patients some of the time. Treatments include using acetylcholinesterase inhibitors or stimulators of ACh release to improve neuromuscular transmission, using immunosuppressive drugs, plasma exchange or plasmapheresis to remove circulating antibodies, and thymectomy to remove thymic myoid cells which might serve as a source of antigenic stimulation. A breakthrough in the characterization of this disease came from the observation that a myasthenia-like disease could be induced in rabbits after injection of an AChR-rich preparation of *Electrophorus* membranes (115). It was subsequently shown that myasthenia could be induced experimentally (experimental autoimmune myasthenia gravis, EAMG) by injecting crude or purified preparations of AChR from apparently any source into a wide variety of test animals (ranging from rodents to primates). The AChR is extremely immunogenic and although animals with EAMG have antibodies that recognize many proteins of the neuromuscular junction, it was shown that only AChRs were capable of disease induction (116). The observations that AChRs and only AChRs could induce EAMG implicated the AChR as a key component in the induction of MG. Although the AChR is clearly involved in myasthenia, given the clinical and immunological heterogeneity that characterizes this disease and evidence that a diversity of genetic factors may influence susceptibility, it is likely that MG is not a single disease (24–26,117–119).

3. Isolation of *Torpedo* AChR cDNAs

Even with the availability of large quantities of material and a number of excellent ligands for studying the AChR, many difficulties have been encountered in its isolation and characterization. Although the subunits are non-covalently associated, they are so tightly associated that they can only be dissociated using conditions that completely denature the polypeptides. Attempts to renature individual subunits into functional AChR complexes were never successful. Thus, with the exception of the α subunit, it was difficult to establish functions for the individual subunits and it was never absolutely certain that the four subunits were those required for a properly functioning receptor-channel. Reconstitution studies using affinity purified *Torpedo* AChRs did establish that the four subunits were sufficient for functioning receptors (120). As will be discussed below, the application of recombinant DNA techniques to the study of this molecule has significantly increased our understanding of its different properties. This approach is helping to unravel the role of the AChR in development, has provided better methods for studying its biosynthesis, has allowed the isolation of novel subunits and of neuronal AChR subunit cDNAs and, in conjunction with the isolation of other ligand-gated receptor subunit cDNAs, has revealed that the AChR is a member of a large superfamily of ligand-gated channels.

Although individual AChR polypeptides are synthesized in cell-free translation systems, attempts to express assembled and functional AChRs in such systems have not been successful (31,121,122). Subunits were not properly glycosylated, they did not assemble into 9S complexes and the BuTx binding function was not reconstituted. The first successful expression of AChRs in a foreign environment was achieved by Sumikawa *et al.* (32) who isolated poly[A]$^+$ mRNA from *Torpedo marmorata* electric organ, microinjected it into *Xenopus* oocytes, and obtained data that indicated that individual subunits were glycosylated and assembled into proper $\alpha_2\beta\gamma\delta$ complexes which were sequestered into oocyte membranes and were capable of binding BuTx. In a later study (123), they showed that the application of ACh to these AChRs resulted in the opening of channels whose ionic permeabilities resembled those of nicotinic AChRs. These results were very important because they demonstrated that the necessary post-translational modifications of subunits and their assembly into pentameric complexes could be properly executed in oocytes. Because heterogeneous RNA was injected into oocytes, the question of whether the four subunits were necessary and sufficient for AChR assembly and function could not be answered. However, these studies were instrumental in the subsequent rapid progress made with cloned AChRs because they established that the oocyte expression system could be used to express functional AChR complexes rapidly and easily.

Simultaneously, around 1980, several groups began to clone the cDNAs for the AChR subunits. At the time these studies were begun, cDNA expression libraries were not available and cloning of a heterologous multisubunit protein complex by DNA-mediated gene transfer techniques was not feasible. Partly due to these considerations and partly because the abundance of AChR proteins in *Torpedo* electric organ suggested that this would be a good source of AChR mRNAs, the groups involved in these early studies prepared cDNA libraries from the mRNA of *Torpedo* electric organ. Furthermore, the availability of *Torpedo* AChR protein sequence data (124) and *Torpedo* subunit-specific antisera (125) suggested two gene cloning strategies. The protein sequence information meant that synthetic oligonucleotides corresponding to known AChR amino acid sequences could be used to directly screen cDNA libraries. The cloned DNAs isolated in this way yielded the sequences for α, β, γ and δ subunits (126 – 129). In an alternative approach, a *Torpedo* electric organ cDNA library was prepared and screened for clones that hybridized to cDNA probes made from *Torpedo* electric organ mRNA but did not hybridize to probes made from *Torpedo* brain (130), liver and/or spleen (131), or brain or liver (132). The positive clones were then used to hybrid-select RNA from electric organ, which was translated *in vitro* using a rabbit reticulocyte lysate system, followed by immunoprecipitation by subunit-specific antisera. This approach yielded cloned DNAs comprising the complete γ (133) and α (134) sequences, and partial β and δ sequences (132). A third approach could be employed once sequence information was published (127) or partial clones were made available (132). Specific oligonucleotide probes were synthesized and partial clones were used to probe a *Torpedo californica* electric organ cDNA library. This very quickly resulted in the isolation of full-length α, β, γ and δ clones (135,136).

When the clones were used to probe blots of the RNA transcripts made in electric organs, transcripts of about 2400 nucleotides (nt) were detected for each of the subunits except δ, for which four transcripts were detected of approximately 5600, 3600, 2400 and 1700 nt (128,132; Claudio, unpublished data). The 5600 nt transcript is the most abundant being present at 20 times the concentration of the others. The use of multiple poly[A]$^+$ addition sites on δ gene transcripts [as has been detected for the dihydrofolate reductase gene (137)] could explain the existence of multiple transcripts, but this seems unlikely since the published *Torpedo* δ sequence (128) contains no potential poly[A]$^+$ addition signals (138) in its 214 bp 3' untranslated region. If there were multiple poly[A]$^+$ sites then one might expect different clonal isolates to contain different sized 3' untranslated regions. This appears not to be the case. Of 12 clones isolated, all have essentially the same size 3' untranslated region (Claudio, unpublished data). The explanation and possible significance of the multiple δ transcripts remains unresolved at this time.

The cDNA clones of the *Torpedo* AChR subunits have subsequently been used as probes to facilitate the isolation of the corresponding sequences from other species (see Section 5) as well as neuronal AChRs (Section 10). I will now discuss how sequence analysis of these clones has allowed predictions to be made regarding the structural and functional properties of domains within polypeptide chains. In order that these predictions might be tested, systems are required for expressing the cloned DNA sequences. These approaches will be discussed in Section 7.

4. Structural predictions based on deduced amino acid sequences from *Torpedo* AChR cDNAs

Before the genes for the four *Torpedo* AChR subunits had been cloned, only partial amino acid sequences data had been obtained for their gene products. These corresponded to amino-terminal residues for the α, β, γ and δ subunits (124), and carboxyl-terminal residues for the α subunit (139). Complete sequences had not been obtained for any of the subunits, partly due to the fact that they contained very hydrophobic peptides composed of stretches of leucine and isoleucine residues which are difficult to separate and sequence. As revealed from the complete amino acid sequences deduced from the sequences of the cloned DNAs, each of the four subunits contains several such regions.

4.1 Size of each subunit

Translation *in vitro* of *Torpedo* mRNA followed by immunoprecipitation with δ-specific antisera and partial amino acid sequence analysis demonstrated that the δ subunit had a processed signal peptide of 21 amino acids (140). Results obtained earlier by the same group (122) were consistent with all four of the subunits containing signal sequences (141). The actual proof that each subunit did indeed contain a signal peptide was not obtained until the four subunits were cloned and sequenced (126–129,133,134). Comparison of the true amino-terminal sequences of each subunit as determined by amino acid sequence analysis (124) with those deduced from DNA sequence analysis revealed that each subunit had a processed signal sequence that varied in length from 17 to 24 amino acids (*Table 1*).

The number of amino acids and calculated molecular weights of each subunit are also shown in *Table 1*. Given that each subunit is glycosylated (142–144), discrepancies between calculated and apparent molecular weight are expected. The calculated and observed molecular weights are within 10% of one another except for the α subunit which differs by about 25%. The presence of adjacent basic residues in the α sequence prompted the hypothesis that this might be a post-translational proteolytic cleavage site (127) which, when cleaved, would yield a calculated molecular weight that more closely approximated the apparent weight. However, amino

Table 1. Primary amino acid sequence analysis of cloned AChRs

Source	Subunit	Signal peptide length (amino acids)	Mature length	Calculated molecular weight (daltons)	Apparent molecular weight (daltons)	Potential asparagine-linked glycosylation sites
Torpedo	α	24	437	50 116	40 000	1
Torpedo	β	24	469	53 681	50 000	1
Torpedo	γ	17	489	56 279	60 000	4
Torpedo	δ	24	501	57 565	65 000	3
Calf	α	20	437	49 897	41 000	1
Calf	β	24	481	55 055	50 000	1
Calf	γ	22	497	55 936	53 000	4
Calf	δ	21	495	56 484		2
Calf	ε	20	471	52 568	56 000	3
Mouse (BC$_3$H-1)	α	20	437	49 898	42 000	1
Mouse	β	23	478	54 660	46 000	1
Mouse	γ	22	497	56 493	48 000	4
Mouse	δ	24	496	57 104	60 000	3
Human	α	20	437	49 694		1
Human	γ	22	495	55 724		4
Chicken	α	19	437	50 150		1
Chicken	γ	22	492	56 484		2
Chicken	δ	18	495	57 215		2
Xenopus	α1b	20	437	52 112	43 000	2
Xenopus	α1a	20	437	50 161	42 000	1
Xenopus	γ	17	493	58 083		1
Xenopus	δ	21	500	59 733		4
Rat brain	α2	27	484	55 480		3
Rat (PC12, brain)	α3	25	474	54 780		3
Rat brain	α4–1	30	599	67 037		4
Rat brain	α4–2	30	600	67 167		4
Rat (PC12, brain)	non-α	28	475	54 336		4
Chicken neuronal	α2	23	503	58 090		3
Chicken neuronal	α4	23	599	68 400		2
Chicken neuronal	non-α	18	473	54 000		2
Drosophila	ALS	20	547			2
Drosophila	ARD	24	497	57 340		1

acid sequence analysis of the carboxyl-terminal end of the α subunit isolated from *Torpedo* electric organ revealed that no such proteolytic processing occurred (139). Discrepancies in calculated and apparent molecular weights are often observed with glycosylated proteins, but one would expect the apparent molecular weight of α (40 kd) to be larger not smaller than the calculated molecular weight (50 kd). The reason for this anomalous migration of the α subunit on SDS polyacrylamide gels has not been determined and may reflect an unusual tertiary structure that is resistant to high concentrations of SDS and reducing reagents, or one that binds more SDS than would be predicted from its calculated molecular weight.

4.2 Potential glycosylation sites

Examination of the deduced primary amino acid sequences also allowed the identification of potential asparagine-linked glycosylation sites (identified by the sequence Asn-X-Ser/Thr). As shown in *Table 1*, the α and β subunits each contain one potential asparagine-linked glycosylation site, the γ contains four (129) or five (133), and the δ contains three. A single base difference between the two reported *Torpedo* γ sequences translates into a serine (33) or proline (129) residue and so this subunit could have either four or five of such potential glycosylation sites. If one uses the four transmembrane-domain model of subunit folding first proposed for the γ subunit (133) (*Figure 1*), then the prediction is that two potential glycosylation sites would be located cytoplasmically and thus there would be either two or three asparagine-linked sites on the γ subunit. Experiments using a wheat germ cell-free protein synthesizing system in the presence of dog pancreas rough microsomal membranes (122) suggested the possibility of three glycosylation sites, whereas the results obtained subsequently from the cloning and sequencing of γ subunits from bovine (145), chicken (146), human (147), mouse (148,149) and *X.laevis* (150) all indicated that the amino acid at position 281 was proline and not serine, strongly favoring the prediction that there were only two external asparagine-linked glycosylation sites in the γ subunit. The results of carbohydrate analysis of *T.californica* AChR are consistent with there being one, one, two and three units of oligosaccharide for α, β, γ and δ respectively (144). Experiments using tunicamycin and endoglycosidase H have recently proved that the subunits do indeed contain one, one, two and three units of asparagine-linked oligosaccharide (151).

An indirect method for determining the location of asparagine-linked glycosylation sites was attempted by Mishina *et al.* (152). Using site-directed mutagenesis techniques, this group destroyed the one potential asparagine-linked site in the α subunit by converting Asn-141 to an aspartic acid residue. RNA was then synthesized *in vitro* from this mutant cDNA, mixed with wild-type β, γ and δ RNAs synthesized *in vitro*, and then microinjected into *X.laevis* oocytes. No [^{125}I]BuTx-binding activity

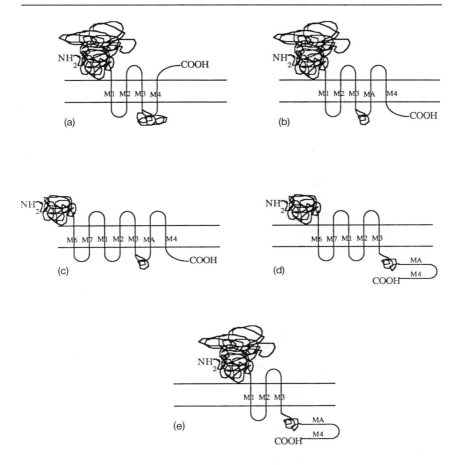

Figure 1. Models for the topology of folding of the AChR subunits in the plane of the membrane. The original model of four transmembrane segments (M1 – M4) is presented in (**a**) and was based solely on hydropathy profiles. An amphipathic helix (MA) was later predicted which suggested five membrane crosses, shown in (**b**). Immunological evidence then suggested models presented in (**c**) (seven crosses) and later (**d**) [five crosses but a different set of crosses than those presented in model (**b**)]. By default, the model presented in (**e**), consisting of only three membrane crosses might also be a possibility.

was detected in solubilized oocytes and no channel activity was seen when ACh was applied to the bathing solution. If Cys-142 were located at the ACh binding site then a charged amino acid at position 141 (aspartate) might severely alter the BuTx- and ACh-binding properties of the receptor (152). However, since we now know that residue 192 is located at or near the ACh site (153; see Section 4.4), it is less likely that the binding site had been altered by this single amino acid change. We further know that each subunit is glycosylated (142–144,151) and that no AChR assembly occurs in the absence of glycosylation (16,17). Thus, when Asn-141 was changed to an aspartic acid residue and no BuTx or ACh binding was detected (152), it is possible that the AChR subunits did not properly assemble into complexes. Whether each of the remaining six asparagine-linked glycosylation sites on the other three subunits has a similarly profound effect on function is not yet known.

4.3 Potential phosphorylation sites

The AChR has been isolated from several sources and shown to be phosphorylated. *In-vitro* studies using *Torpedo* membrane fragments have demonstrated that the AChR can be phosphorylated by at least three different protein kinases: cAMP-dependent protein kinase (PKA) which phosphorylates the γ and δ subunits (154), protein kinase C (PKC) which phosphorylates the δ (155,156) and possibly the α subunit (155), and an endogenous tyrosine kinase which phosphorylates the β, γ and δ subunits (157). Having the complete amino acid sequence for each subunit has greatly aided investigators in identifying the specific phosphorylation sites of PKA and PKC on the subunits (155–158). The role of phosphorylation in modulating AChR electrophysiological properties has long been debated (reviewed in refs 20,158). Reconstitution studies have shown that PKA (159) and tyrosine kinase (160) increase the rate of desensitization of *Torpedo* AChRs. In other studies using mammalian cell lines or primary cultures, stimulation of PKA increased the rate of desensitization (161–163) and stimulation of PKC decreased activity (164). In addition to altering electrophysiological properties, stimulation of different kinases alters the number of cell surface receptors. Stimulation of PKA was shown to increase the number of cell surface receptors (165–167) while PKC was shown to cause a decrease in their number (167). In chick muscle, the δ subunit was found to be phosphorylated before as well as after assembly suggesting a role of phosphorylation in the assembly process (168). Differences in the pattern and degree of phosphorylation of the γ and δ subunits were observed for *Torpedo* receptors stably expressed in mouse fibroblasts depending on whether the subunits were assembled into complexes on the cell surface, assembled into complexes and located intracellularly, or unassembled (169). The pattern and degree of phosphorylation in these three different subunit pools could be altered by stimulation of PKA suggesting that changes in surface expression result from a direct phosphorylation of the receptor (169).

4.4 Location of the ACh binding site

A number of studies have established that the ACh binding site is located on the α subunit (1–7). Karlin and co-workers synthesized an alkylating compound, MBTA (maleimidobenzyltrimethylammonium) (170), which competed with ACh for binding to the AChR. When the AChR was mildly reduced and affinity-alkylated with [^{3}H]MBTA, only the α subunit was labeled. They also demonstrated that there was a readily reducible cysteine disulfide bond within 1 nm of the ACh site which, when reduced, drastically altered the binding properties of the AChR (1,170). Thus from biochemical analysis, it is known that the ACh binding site is located on the α subunit at or near a cysteine residue.

In the paper describing the first complete sequence of the α subunit, Noda et al. (127) attempted to predict the location of this important site from primary amino acid sequence data. Two criteria for identifying the binding site are that a cysteine residue must be present and that it must be located extracellularly. Seven cysteines are present in the α subunit (at positions 128, 142, 192, 193, 222, 412 and 418). Three of these (at positions 222, 412, 418) are located in very hydrophobic domains which the Kyte and Doolittle (171) hydropathy plot predicted to be transmembrane segments (see Section 4.5 and *Figure 1*). These cysteine residues were thus unlikely candidates for the ACh site. Of the remaining four cysteine residues, Cys-128 and -142 were hypothesized (93) to be the ACh site for the following reasons:

(i) two negatively charged residues (Glu-129 and Asp-138) located between Cys-128 and -142 might serve as a negative subsite (170) to attract the positively charged quaternary ammonium group of ACh;

(ii) the only potential asparagine-linked glycosylation site on α is located on Asn-141 and if α is glycosylated at this residue, then this area would be extracellular;

(iii) this area is predicted to contain β-sheet structures, suggesting a tightly packed region that might stabilize a specific conformation (possibly required for forming the ACh binding site).

Although this model (127) was very attractive because it accommodated much of the biochemical data, it was later demonstrated that Cys-192 and -193, and not Cys-128 and -142, were located at or very near the ACh binding site. Kao et al. (153) used [^{3}H]MBTA to affinity-alkylate the AChR and subsequently isolated and sequenced the alkylated [^{3}H]-labeled CNBr peptide. They demonstrated that Cys-192 (and to a lesser extent Cys-193) was labeled. In a subsequent study (172), they determined that disulfide linkages occurred between Cys-192 and -193 and between Cys-128 and -142.

4.5 Topology of folding of subunits in the plane of the membrane

One of the biggest controversies in the AChR field relates to the topology of folding of the subunits in the plane of the membrane. The hydropathy profile obtained for the deduced amino acid sequence of the *Torpedo* γ subunit indicated four regions that were long enough and hydrophobic enough to span the lipid bilayer (133; *Figure 1a*). Subsequent hydropathy analyses of the sequences of the α (134) and the α, β, γ and δ (129) subunits indicated the same four putative membrane-spanning domains (denoted M1 – M4). Because the function of the membrane-spanning domains of the AChR is not merely to anchor the protein in the membrane but also to form a cation-selective channel, it was possible that the amino acids which lined the pore of the channel might contain charged and/or polar side-chain residues. Because M1 – M4 do not contain any charged residues and there are only a few polar residues in each of these domains, two groups searched for regions of the molecule that might be strongly α-helical and amphipathic in nature (173,174). Such a region was located between M3 and M4 in all four subunits (*Figure 1b*), termed MA (A for amphipathic) or M5.

The finding that there was a high degree of identity between pairs of subunits [ranging from 36 to 49% and an overall identity of 19% amongst all four subunits (30)], together with the finding from hydropathy analyses that each of the subunits appeared to weave through the plasma membrane in a similar fashion (129), suggested that if the topology of folding for one of the subunits could be determined then the results would be directly applicable to the other subunits. If there were four transmembrane domains, then the N- and C-termini would be located on the same side of the membrane. If there were five transmembrane domains, then the N- and C-termini would be located on opposite sides of the membrane. From biochemical studies on the *Torpedo* δ subunit (175) and from *in-vitro* translation studies which demonstrated that the δ subunit contained a processed signal peptide (140), it was concluded that the four AChR subunits had their N-termini located extracellularly. The 'four transmembrane domain' model (*Figure 1a*) thus predicts that the C-terminus is on the extracellular surface and the 'five transmembrane domain' model (*Figure 1b*) predicts that it is on the cytoplasmic side.

4.5.1 *Predictions using antibodies*

The initial controversy concerning subunit topology centered around the 'four' versus the 'five' transmembrane domain models, but quickly expanded to include other possible models. The first studies were designed to locate the C-terminus of at least one of the subunits and thereby distinguish between these two models. Studies using monoclonal antibodies directed against C-termini of different subunits or directed against synthetic peptides of different subunit C-termini all indicated that the C-terminus was located cytoplasmically (176 – 178) suggesting that there were five membrane crosses (*Figure 1b*).

A more extensive analysis with other antibodies, however, complicated the picture. Two monoclonal antibodies were subsequently identified which were believed to be directed against an epitope located on the α subunit around position 152 – 159 which, according to biochemical binding studies, appeared to be a cytoplasmic domain. Because the asparagine-linked glycosylation site is located at position 141 and therefore has an extracellular location, and the ACh binding site is located at position 193 and therefore also has an extracellular location, if position 152–159 is cytoplasmic, the protein must cross the membrane two additional times before the M1 domain, making a total of seven (M6 and M7) transmembrane domains (*Figure 1c*) (179). The M6 segment in the seven membrane domain model crosses the membrane with only ten amino acids (residues 142–151), an insufficient number to form an α-helix. It must therefore cross in some extended conformation. The M7 domain in this model could form an α-helical amphipathic segment which spans the membrane somewhere between residues 160 and 185 (179).

Continued analysis by competition binding studies using monoclonal antibodies directed against several epitopes of the MA domain and regions closely flanking it all indicated that the monoclonals were directed against cytoplasmic epitopes (180,181). The results obtained from these antibody studies indicated that both MA and the C-terminus were located cytoplasmically, which meant that M4 was also cytoplasmic. A new model of five transmembrane domains was thus proposed in which the subunits were composed of M6, M7, M1, M2 and M3 domains (*Figure 1d*) (180).

There are several difficulties with trying to predict polypeptide folding patterns from experiments using antibodies. One approach involves making specific synthetic peptides for use as immunogens, the resulting antibodies being used to try to locate the epitope on a protein in its native conformation. Problems can arise because the conformation of the peptide and the protein are different, resulting in antibody that either does not bind or, much worse, binds to the wrong site. In another approach, antibodies are prepared against native AChRs and then are screened to determine their subunit specificity. Unfortunately, the AChR subunits cannot be dissociated without denaturing the polypeptides. The common method for determining antibody subunit specificity is by immunoblotting, which again requires that the antibodies recognize the same epitope in native and denatured states. One other difficulty in using antibodies to determine protein shape can be attributed to the inaccessibility of sites on the AChR to antibodies without prior treatment by chemicals or detergents. This inaccessibility problem can be attributed to polypeptide folding, tight packing of AChRs, or association of other proteins with the AChR such as the cytoplasmic muscle-specific 43 kd protein (103).

Preparations of AChR-rich membrane vesicles can be obtained from *Torpedo* electric organ in which more than 95% are sealed right-side-out. Triton X-100 treatment will completely solubilize the vesicles yet the receptor complexes remain intact. Saponin treatment merely makes holes in the membranes without solubilizing them and lithium diiodosalicylate (LIS) treatment both disrupts the vesicles and removes proteins such as the 43 kd protein. These different permeabilizing treatments have been used to try to locate the subunit N- and C-termini. Difficulties with antibody binding have been reported by several groups using antibodies directed against the N-termini of the different subunits (176,182). None of the subunit N-terminal-specific antibodies have been capable of binding to native AChRs although each is capable of binding to denatured polypeptides. The interpretation of these results cannot be that the antibodies are not binding to native AChRs because the N termini are located cytoplasmically, since it is known that the N terminus is extracellular (140,175). The results are best explained by assuming that the N terminus is merely inaccessible to antibodies when the receptor is in its native conformation. In contrast, when antibodies directed against the C termini of each subunit did not bind to AChR without prior treatment of the AChR with Triton X-100, saponin or LIS, it was assumed that the C terminus was cytoplasmic (176,177). An alternative explanation is that the C terminus is extracellular but inaccessible to antibodies in its native conformation.

The inaccessibility problem was elegantly addressed by LaRochelle *et al.* (183) who used a double gold-labeled antibody technique and electron microscopy. First a defined epitope such as the BuTx binding site was tagged by incubating a preparation of LIS- or saponin-treated AChRs with BuTx, followed by anti-BuTx antibodies which were coupled to 15 nm colloidal gold particles. A second antibody of unknown topological specificity was coupled to a colloidal gold particle of a different size (5 nm). By electron microscopy, one can easily distinguish the two sizes of gold particles and can determine if the unknown antibody is binding to the same side of the membrane as BuTx (extracellular) or to the opposite side (cytoplasmic). One major advantage of this technique is that one always has a point of reference (e.g. the extracellular BuTx binding site) and so, no matter how severely the membrane has been disrupted with chemical or detergent treatments, the binding orientation of the uncharacterized antibody can always be determined. The technique also provides a good internal standard for antibody binding. The distribution and degree of labeling by both the reference antibody and the uncharacterized antibody should be similar.

The results obtained from double gold labeling of a large number of different MA epitopes have convincingly demonstrated that MA is not a membrane-spanning domain and that it is located cytoplasmically (180, 181). Because so few antibodies have been found that react with the C-terminus, the data that this site is also located cytoplasmically are less convincing, and the data that support the existence of an M6 and an M7 domain are less convincing still. The only evidence for an M6 and M7 domain is that two monoclonal antibodies directed against a synthetic peptide of the α subunit (residues 152–167) bind the AChR but only after it has been treated with LIS or saponin (179). The experiments describing the specificity of this monoclonal antibody have all been biochemical and no double gold labeling experiments have yet been reported. It remains a formal possibility (by analogy with the binding results obtained with anti-N-terminal antibodies) that the epitope could be extracellularly located but simply inaccessible to antibodies without prior treatment by reagents that disrupt the membrane. Other immunological evidence suggests that M7 is not a transmembrane domain, and that it may be located extracellularly. Using a collection of 18 consecutive and overlapping 16- or 17-residue synthetic peptides covering residues 1–210 of the *Torpedo* α subunit, Mulac-Jericevic *et al.* (184) determined the most antigenic regions along this segment. Eight regions along the 210 amino acid stretch were found to be antigenic, with five of them being immunodominant. One of these immunodominant regions had boundaries between residues 172 and 182. The M7 domain was predicted to consist of residues 160–185. Because antiserum directed against membrane-bound AChRs binds to the synthetic peptide 170–186, this suggests that the region is not transmembrane and that it is probably located extracellularly. If there are no M6 and M7 domains, and if both MA and M4 are located cytoplasmically, then yet another model could be proposed. Such a model would contain only the three membrane domains, M1, M2 and M3 (*Figure 1e*).

4.5.2 Predictions using site-directed mutagenesis

Mishina *et al.* (152) have made a series of deletion mutations in the α subunit of *Torpedo* in order to determine the role of particular sequences of amino acids in AChR channel function. Some of their results could also be used to infer whether the deleted regions are simply membrane-spanning and not necessarily channel-lining. The group systematically deleted M1, M2, M3, M4 and several regions of the MA domains of the α subunit. Transcripts were prepared *in vitro* from the mutated α and wild-type β, γ and δ subunit cDNAs, and were injected into oocytes which were subsequently tested for AChR function with two assays. The oocytes were solubilized and incubated with [^{125}I]BuTx to see if subunits assembled into complexes, or voltage-clamped in order to record the current after application of ACh to their bathing medium. If any of the M1, M2, M3 or M4 domains were deleted, BuTx binding activity was retained, but no ACh-activated channel activity was detected. The BuTx binding was reduced to 0.5% of normal for M1, 17% for M2, 8% for M3 and 23% for M4 but, as there was some residual binding, presumably the mutations were not so severe that they prevented assembly. However, BuTx binding to the cell surface was not tested, and it is not known if the correct AChR complexes were formed since the physicochemical properties of the binding material were not analyzed. Ambiguous results were obtained with deletions in the MA domain. Six different deletions resulted in all of the oocytes responding with 29–45% of normal BuTx binding function. In contrast, three of the deletions resulted in no detectable ACh response, whereas one of the deletions (which removed the entire MA) resulted in 100% of the oocytes responding to ACh but with depolarizations of only 3% of the control values. One interpretation of the deletion mutation experiments is that M1, M2, M3 and M4 are each membrane-spanning domains. If any one of these domains is removed, subunits are not properly assembled, the complex is not inserted into the oocyte membrane, and no detectable response to ACh is observed. Because removal of MA resulted in functional (albeit poor) cell surface AChRs, one interpretation of this result is that this domain is not membrane-spanning. Mishina et al. (152) reached the opposite conclusion, namely that MA was membrane-spanning and probably formed the inner wall of the channel. Depending on one's interpretation of these data, the study either favors the model presented in Figure 1a or in 1b.

4.5.3 Predictions using electron image analysis

One method of establishing the number of times the peptide chain crosses the membrane would be with X-ray crystallographic studies. Unfortunately, the AChR has not been successfully crystallized, and so this has not been possible. Two-dimensional crystalline arrays of AChR in membrane have been analyzed by electron microscopy (73). The resulting 2.5 nm resolution model is compatible with five membrane-spanning segments per subunit. However, these two-dimensional crystals contain the cytoplasmic receptor-associated '43 kd' protein (104), the presence of which could complicate the interpretation (185). Thus, although the results are compatible with five crosses, they could be reconciled with four.

4.5.4 Predictions using a biochemical approach

A biochemical approach was also taken to try to determine whether the polypeptides spanned the membrane an even or an odd number of times. AChR isolated from *Torpedo* but not from other species can be found in two forms: in addition to the 'monomeric' 250 kd complex, there is also a 'dimeric' 500 kd complex composed of two monomers covalently cross-linked by cysteine disulfide bonds between δ subunits. Examination of the amino acid sequences of δ subunits from *Torpedo* (128), mouse (186), chicken (146), calf (187) and *Xenopus* (150) reveals a high degree of conservation of cysteine residues. Of the 5 – 8 cysteine residues found in each species, four are conserved in all five species and another is found in all species except *Xenopus*. The only cysteine residue found uniquely in one species is that in *Torpedo* and is located at the extreme C-terminal end (the penultimate residue). It is intriguing to propose that the disulfide bond formed between δ subunits that results in the expression of AChR dimers only in *Torpedo* is accomplished with this unique cysteine residue.

Similar approaches were taken by two groups to attempt to determine if the disulfide bridge between subunits was located on the extracellular or the cytoplasmic surface of the membrane (188 – 190). Sealed channel-containing vesicles were prepared and reducing agents that are thought to be membrane-impermeable were used to disrupt the disulfide bridge. The results using glutathione (188,190) and β-mercaptoethane sulfonic acid (189) indicated that the disulfide responsible for *Torpedo* δ dimers was located on the extracellular side of the membrane. If the membrane vesicle remained impermeant to the applied reducing agents, as suggested by the results of both experiments, then the cysteine residue responsible for formation of δ–δ dimers is located on an extracellular domain. The question now becomes: which cysteine residue is involved in disulfide bond formation? Neither study actually isolated the fragments containing the disulfide. However, earlier proteolysis experiments by Wennogle et al. (191) using *T.marmorata* membrane vesicles indicated that the disulfide was located on the C-terminal side of M3. Although mouse, chick, calf and *Xenopus* each have one or two cysteine residues in this region, there is only one in *T.californica*, the penultimate Cys-500. Using the same strategy as described above, DiPaola et al. (192) also obtained results indicating that the disulfide cross-link between δ subunits was located on the extracellular surface of the membrane. However, in this study the cystinyl residue involved in formation of the δ–δ dimer was identified. The sulfhydryls involved in dimer formation were alkylated with [^{3}H]*N*-ethylmaleimide, then identified by amino acid analysis and sequence analysis of the C-terminal CNBr fragment. In addition, the CNBr fragment containing the unreduced δ–δ dimer was also analyzed. The results from both studies indicated that the cysteine residue involved in δ–δ dimer formation was Cys-500.

The results from these studies indicate that the C-terminus of δ is on the extracellular surface of the membrane, implying that there are an even number of membrane crosses and so favoring model 1 (*Figure 1a*). Because of the similarity in hydropathy profiles among the four subunits, it is presumed that each subunit probably spans the membrane four times. The question of the number of transmembrane domains for the AChR subunits is still somewhat controversial. I feel, however, that the data have accumulated on the side of the original model of four crosses (133). Interestingly, when the amino acid sequences of two other recently cloned ligand-gated receptor-channels [the glycine receptor (34) and the GABA receptor (35)] were subjected to hydropathy analysis, they both appeared to have four transmembrane domains (see Section 11.1.3), remarkably similar to the four observed for the AChR subunits.

5. Isolation of nicotinic AChR cDNAs from other species

Once the primary amino acid sequences for the *Torpedo* AChR subunits had been deduced, predictions about the topology of folding of the

Table 2. (Muscle-like) AChR subunit similarities

	Torpedo				Bovine					Murine				Human		Chicken			Xenopus			Rat (PC12, brain)	Drosophila
	α	β	γ	δ	α	β	γ	δ	ε	α	β	γ	δ	α	γ	α	γ	δ	α1b	γ	δ	α₃	Non-α
Torpedo																							
α		41	36	36	79	38	33	38	31	80		33	36	80	33		36	37	77	33	33	47	44
β	41		41	40	42	57	42	43	39	42	59	41	41	40	40		43	42	40	41	40	38	42
γ	36	41		49	36	41	54	48	57	36		54	48	35	55		62	50	35	59	50	34	34
δ	36	40	49		36	39	45	59	45	36		45	59	34	43		49	61	37	47	61	38	37
Bovine																							
α	79	42	36	36		40	35	38	31	96	33		37	97	33		36	37				49	42
β	38	57	41	39	40		42	42	40		90	41	41	38	41		42	41				37	41
γ	33	42	54	45	35	42		49	53	35		90	48	32	92		66	48				35	33
δ	38	43	48	59	38	42	49		45	38		47	87	35	46		50	68				39	38
ε	31	39	57	45	31	40	53	45		35		51	46	30	53		54	44				35	34

	Torpedo α β γ δ	Murine α β γ δ	Human α γ	Chicken α γ δ	Xenopus α1b γ δ	Rat α3	Drosophila Non-α
Murine (BC₃H-1)							
α	80 42 36 36	96 35 38 35	40 33 36	95	86 36 38	51	
β	59	33 90	40				
γ	33 41 54 45	41 90 47 51	33 47	90	68 46	36	
δ	36 41 48 59	37 41 48 87 46	36 47	33	50 68	39	
Human							
α	80 40 35 34	97 38 32 35 30	95	33			
γ	33 40 55 43	33 41 92 46 53		90			
Chicken							
α	36 43 62 49	36 42 66 50 54	86				
γ			36		49	37	
δ	37 42 50 61	37 41 48 68 44	38		49	38	
Xenopus							
α1b	77 40 35 37						
γ	33 41 59 47						
δ	33 40 50 61						
Rat (PC12)							
α3	47 38 34 38	49 37 35 39 35	51	36 39		37 38	46
Drosophila							
Non-α	44 42 34 37	42 41 33 38 34					46

Numbers represent per cent identities or similarities (identity plus conservative substitutions).

individual subunits, the location of the ACh binding site, and the amino acids that formed the inner lining of the channel quickly began to amass. *Torpedo* cDNAs were used to isolate AChR subunits from other species and the subsequent sequencing of these genes and cDNAs has contributed substantially to the number of predictions being made about structure and function.

5.1 Isolation of AChR cDNAs

The α, β, γ and δ AChR subunit cDNAs from calf have been isolated using the poly[A]$^+$ fraction of mRNA obtained from newborn or fetal bovine skeletal muscle tissue as the starting material and the cloned cDNAs from *Torpedo* to probe the cDNA libraries (145,187,193,194). A similar strategy was used in order to isolate mouse subunit cDNA clones from the murine BC$_3$H-1 (196) muscle cell line (148,149,186,196–199). *Torpedo* subunit cDNAs were similarly used to isolate α, γ and δ subunit clones from *X.laevis* embryonic cDNA libraries (150). Two unexpected discoveries have resulted from the screening of different libraries (see Section 6). A novel γ-like subunit, ϵ, was retrieved from a fetal bovine library. In addition, a second muscle α clone was obtained from a *Xenopus* embryonic library (200).

5.2 Sequence similarities

The extent of sequence identity among the four subunits of *Torpedo* AChR is 19%. Pair-wise identities between subunits range from 36% for the α/γ or α/δ pair to 49% for the γ/δ pair (see *Table 2* and ref. 30). The most highly conserved subunit across species is the α subunit which ranges from 79% identity between *Torpedo* and bovine to 96% among the mammalian species (mouse, bovine and human). Although identical segments can be found across the entire length of the different subunits, the most highly conserved sequences generally fall within the four membrane-spanning domains (M1–M4) and the sequences around the acetylcholine binding site (Cys-192 and -193). A comparison of the amino acids within M1, M2 and M3 reveals that they are identical between human and bovine α sequences and that 19 of 20 amino acids are identical between the M4 sequences of these species. The overall similarity (identity plus conservative substitutions) between human and bovine α sequences is 97%. The per cent identity between bovine and *Torpedo* α sequences for the M1–M4 domains is 96, 100, 88 and 80%, respectively. The overall identity between bovine and *Torpedo* α sequences is 79%. A comparison of the M1–M4 domains among α, β, γ and δ subunits from human (α, γ), bovine (α, β, γ, δ) and *Torpedo* (α, β, γ, δ) indicates identities of 26, 26, 17 and 15% for M1–M4, respectively. Similar comparisons of the four subunits from these three species for M6, M7 and MA indicates identities of 16, 11 and 4%, respectively. That M1 and M2 are usually the most highly conserved putative transmembrane domains could indicate

Table 3. (Neuronal) AChR subunit similarities

		Rodent					Chicken					Drosophila	
		α1	α2	α3	α4	Non-α	α1	α2	α3	α4	Non-α	α	Non-α
Mouse muscle	α1		49	47	50	43						43	
Rat brain	α2	49		57	67	50						40	
Rat (PC12, brain)	α3	47	57		57	49			84			43	
Rat brain	α4	50	67	57		53				73		44	
Rat (PC12, brain)	non-α	43	50	49	53						43		
Chicken	α1							50	53	51	44		
Chicken	α2						50		65	80	55		
Chicken	α3			84			53	65		66	54		
Chicken	α4				73		51	80	66		55	44	
Chicken	non-α					43	44	55	54	55			
Drosophila	α										44		42
Drosophila	non-α											42	

Numbers represent per cent identities or similarities (identity plus conservative substitutions).

that they perform some important structural or functional role. Indeed, the possible involvement of these domains in the formation of the ion channel has been suggested by several laboratories (see Section 5.3).

In addition to highly conserved regions of the molecule, there are several invariant residues. In the α subunit (but no other subunit) of all species (including the neuronal subunits), there is a pair of cysteine residues located at positions 192 and 193 (*Torpedo* numbering). This is to be expected because Cys-192 and -193 have been identified as being located at or near the ACh binding site (153). As will be discussed in Section 10, the presence of these two cysteine residues is used as one of the criteria for identifying neuronal AChR α subunits where the identity between skeletal and neuronal α subunits is only about 50% (*Tables 2* and *3*). Muscle-like α subunits from the six species isolated thus far are each composed of 437 amino acids. In all subunits of all species there is a proline residue in the middle of M1, cysteine residues at positions 128 and 142, and a potential asparagine-linked glycosylation site at position 141 (*Torpedo* α numbering).

5.3 Location of channel-lining sequences

An extremely interesting structural and functional domain of the AChR is the actual pore of the AChR-channel. It should be illuminating to determine the amino acids that are located at the entrance to the channel and that form the lining of the channel in terms of better understanding ion permeation, channel selectivity, and how ligand binding is coupled to gating. The subunits are arranged about a central pit and it is thought that probably one transmembrane domain from each subunit contributes to the lining of the channel. Two models have been put forward in which it is believed that the channel domain is composed of several positively and negatively charged residues: M5 (173,174) or M7 (180). Other models have been proposed in which no charges line the channel but in which polar residues are, however, present (133,134).

Given the similarity in channel properties across species, it is not unreasonable to postulate that certain structures that contribute to a specific function might also be highly conserved across species (33). Structural studies have shown that the open pore of the AChR-channel is approximately 12 nm in length and the extracellular entrance of the channel is about 2.5 nm in diameter (73,201). From organic cation permeation studies, however, it was demonstrated that the channel must be constricted at some point to a minimum diameter of only 0.64 nm (74,86,202). The channel is believed to be a water-filled pore; a large portion of the wall of which is not lined by high-field strength charges (86). Other evidence suggests that there is a net negative charge within the channel (202), probably located in the wide vestibule (203).

5.3.1 Evidence favoring either M1 or M2 as channel-lining sequences

Using the single-channel patch-clamp technique to study the monovalent and divalent cation permeation characteristics of AChRs in rat myotubes, Dani and Eisenman (204) obtained results that were consistent with a channel having a long, wide, multiply occupied vestibule and a short, selective, singly occupied vestibule, with a small amount of net negative charge within the pore. Of the putative membrane-spanning domains, the highly conserved M1 and M2 sequences seemed the more likely candidates for lining the pore (204,205). Of these two sequences, M1 is an attractive candidate because it contains a highly conserved proline residue, present in all subunits of all species. Proline residues break α-helical structures exposing carbonyl dipoles which could serve to coordinate an ion in a binding site (206). Proline residues also cause a bend in the polypeptide backbone which might serve as the site of constriction of the channel. Because positively charged ions flow through the AChR, one might expect the presence of negative charges at the mouth of the channel to serve as an attractant of ions to the pore (205). M2, on the other hand, is an appealing candidate because there are several negatively charged residues between M2 and M3 which might serve this function. The charge distribution contributed by the regions between M2 and M3 from $\alpha_2\beta\gamma\delta$ AChR subunits from several species is as follows. *Torpedo* has six negative charges and eight positive charges: a net positive charge of two. Calf has five negative and ten positive: a net positive charge of five, and mouse has five negative and nine positive: a net positive charge of four. As will be described in more detail in Section 8.8.2, studies with chimeric δ subunits or point mutations in individual subunits have indicated that charged residues in the regions between M1 and M2 or between M2 and M3 can indeed influence ion transport (207,208). The structural predictions made concerning the transmembrane folding pattern (Section 11.1.3) of the recently cloned and sequenced GABA receptor-channel are also of interest (35). When the deduced amino acid sequence was subjected to hydropathy analysis (171,209,210), four transmembrane domains were also predicted for this receptor. Interestingly, the GABA receptor (which permeates negatively charged chloride ions) has a net positive charge of 8 (assuming an $\alpha_2\beta_2$ stoichiometry) in the region between M2 and M3 (ten positively charged residues and only two negative residues).

5.3.2 Evidence favoring M1 as channel-lining sequences

A number of local anesthetics including quinacrine azide can be used to label the high-affinity non-competitive inhibitor (NCI) site. Evidence for M1 lining the channel has been obtained using quinacrine azide to photolabel the NCI site (85,211). This compound acts as a voltage-dependent, open-channel blocker and promotes desensitization (79,85,212, 213). When [^{3}H]quinacrine azide and AChR-rich membrane fragments from *Torpedo* were irradiated in the presence of ACh (but not competitive antagonists), the AChR was transiently susceptible to photolabeling and the 'specific' labeling was inhibited by various NCIs (213,214). The labeling was confined to the α and β subunits and the site of labeling on the α subunit was localized to the CNBr fragment containing M1 sequences (85).

5.3.3 Evidence favoring M2 as channel-lining sequences

The local anesthetic triphenylmethylphosphonium (TPMP$^+$) has been used to localize the NCI site (77). TPMP$^+$ is a lipophilic cation which binds predominantly to one site on the AChR and blocks ion flux through the channel without inhibiting binding of ACh (81). In the absence of any effectors, α, β and δ subunits were labeled. The yield of labeled subunits was low, only about 1%. Gel-isolated subunits were cleaved into fragments with CNBr or trypsin and subjected to sequence analysis. Only the radioactivity in each cycle was determined, but it was inferred that the labeled residues were Ser-262 of δ, Ser-254 of β and Ser-248 of α. Each of these residues resides within the M2 domain, suggesting that M2 from each subunit forms the inner wall of the cation channel.

Similar experiments with another local anesthetic, chlorpromazine (CPZ), also gave results implicating M2 as a channel component (78). It has been shown that [^{3}H]CPZ labels all four subunit chains of *Torpedo* in the presence of agonists (76,215) suggesting that each subunit contributes to the NCI site and that the site would be located in the axis of 5-fold symmetry composed of at least one transmembrane domain from each subunit. AChR-rich membrane fragments were first incubated with [^{3}H]CPZ in the presence of concentrations of agonist (2 mM carbamylcholine) which would place the AChR in a desensitized (channel closed) state and then irradiated with 254 nm light. The δ subunit was isolated on SDS gels, subjected to tryptic and CNBr digestions, and then sequenced. Again, only the radioactivity in each cycle was determined but it was inferred that Ser-262 of the δ subunit was labeled, indicating that M2 might be one of the channel components (78).

5.3.4 Site-directed mutagenesis

The deletion *in vitro* of sequences encoding domains M1 – M4 (152) did not help to establish which residues lined the channel because they resulted in a non-functional AChR. Of the four deletions, however, deletion of M1 resulted in the least amount of BuTx binding activity (0.5% compared with 17%, 8% and 23%, respectively for M2 – M4) possibly indicating a more important role for this domain beyond that of merely anchoring the subunit in the membrane. On the other hand, it could simply indicate that proper folding of the first membrane-spanning domain is the most critical for establishing a toxin-binding conformation and is unrelated to channel formation. In a later study by this group, the M1 – M4 and MA domains of the *Torpedo* α subunit were each replaced by synthetic double-stranded oligonucleotides corresponding to the hydrophobic transmembrane domain of either the vesicular stomatitis virus glycoprotein or the human interleukin-2 receptor (216). Interestingly, only replacement of the M4 domain resulted in functional channels (whole-cell membrane currents were recorded after activation by bath-applied 1 μM ACh). In no case did substitutions of *Torpedo* α M1, M2, M3 or MA result in functional channels. In all but one case, there was also no detectable cell surface BuTx-binding material [when M2 was replaced by the transmembrane segment of the interleukin-2 receptor (but not the vesicular stomatitis virus glycoprotein), 8% of wild-type BuTx-binding material was detected]. Because deletions or substitutions of M1, M2 or M3 resulted in non-functional channels it was not possible to conclude which domains lined the channel.

5.3.5 Conclusions regarding the channel-lining sequence
The amino acid sequences lining the pore of the AChR channel have not yet been identified, although amino acids that are capable of affecting ion permeation, and thus might be located at the entrance to the channel, have been identified (see Section 8.2.2). Labeling studies using local anesthetic binding have thus far yielded different results. There are several inherent difficulties with studies using local anesthetics because the compounds often bind with different affinities and bind to different subunits depending on the functional state of the receptor. In addition, the yields of labeled fragment are very low and so one cannot readily assess whether the binding that is observed is the physiologically significant binding site. Further, when NCIs are believed to function by physically blocking the open channel, the results of labeling studies conducted on desensitized AChRs (when the channel is closed) are perplexing. Although these sorts of experiments are not easy to conduct they are invaluable for defining and identifying the important agonist, antagonist and non-competitive inhibitor binding sites of the AChR. Furthermore, when either directed or random mutagenesis is applied to trying to define the physiologically relevant ligand binding sites, these approaches will ultimately have to be coupled to such labeling studies.

6. Isolation of novel AChR subunits

During the process of screening various libraries for skeletal muscle AChR subunit cDNAs, two novel findings resulted. A γ-like subunit was isolated (ϵ) from calf and later rat, and a second α cDNA was isolated from *Xenopus*.

6.1 An ε subunit

During the initial isolation of the full-length calf γ clone, a partial calf γ cDNA clone was isolated from a cDNA library prepared from newborn calf muscle poly[A]$^+$ RNA using a *T.californica* γ probe. The partial calf γ cDNA was used to screen a 4-month-old fetal calf cDNA library, which resulted in the eventual isolation of a full-length calf γ cDNA clone (145) and a full-length novel subunit clone (217). The novel subunit (ε) was 53% identical with calf γ and 45% identical with calf δ suggesting that ε was a γ-like subunit. Proof of the identity of ε came, however, from 'mix-and-match' expression studies. In the initial studies, mRNAs were prepared *in vitro* using each of the four *Torpedo* cDNAs and γ and ε calf cDNAs as templates. Microinjection of *Torpedo* α, β, δ, plus calf γ or ε mRNAS into *Xenopus* oocytes resulted in expression of functional AChR-channels, thereby demonstrating that calf γ and ε could substitute for the *Torpedo* γ subunit, and further indicating that ε was a γ-like subunit (217).

A more extensive physiological characterization of these clones indicated that AChRs composed of calf α, β, γ and δ had a conductance reminiscent of fetal calf AChR-channels (low-conductance) whereas AChRs composed of α, β, ε and δ had a conductance reminiscent of adult bovine AChR-channels (high-conductance) (218). In the initial analysis of the ε gene transcripts, the ε probe was found to hybridize to 4-month-old fetal calf RNA but not to 7-month-old fetal calf, 9-month-old fetal calf, newborn calf or adult bovine diaphragm RNA (217). These results were inconsistent with the channel properties which indicated that ε was an adult form. In a later study (218) the results of the RNA analysis were reversed and a consistent picture of RNA expression and channel properties was presented (ε RNA was only detected after birth and α, β, ε and δ channels had conductances similar to adult AChRs). These very exciting results suggested that there might be a developmental switch occurring with AChR ε and γ subunits. It is still a little puzzling, however, that the ε clone was isolated from a cDNA library prepared from RNA isolated from 4-month-old fetal skeletal muscle.

Witzemann *et al.* (219) have recently isolated two partial rat genomic clones apparently encoding rat γ and ε subunits. Investigation of their expression using RNA isolated from innervated, denervated and reinnervated rat muscle demonstrated clear differences in the expression of these two subunits. The ε subunit expression was relatively independent of whether the muscle was innervated, denervated or reinnervated. Upon denervation, the level of ε RNA only increased 2- to 3-fold. In contrast, the γ subunit expression was strongly affected by the state of innervation of the muscle: γ RNA was undetectable in innervated and reinnervated muscle but increased to levels 10-fold higher than ε RNA in the denervated state. The expression of these two subunit RNAs was also coincidental with expression of the high-conductance channel (seen in innervated muscle) and the low-conductance channel (seen in denervated muscle), strongly supporting the notion of a developmental switch occurring via a switch in γ and ϵ subunits.

6.2 Two *Xenopus* α subunits

Baldwin et al. (150) isolated α, γ and δ subunit clones from stage 17 and stage 22 – 24 embryo *X.laevis* cDNA libraries. Hartman and Claudio (203) isolated a second α subunit cDNA from the stage 17 library. Both cDNA clones were identified as being α subunits because of the presence of cysteine residues at positions 192 and 193. They were further identified as being muscle-like (not neuronal-like) because they were composed of 437 amino acids and contained one potential asparagine-linked glycosylation site at residue 141 (see Sections 5.2 and 10, and *Table 1*). Two other observations also indicated the clones were muscle-like. When cRNA was prepared *in vitro* from the clones and expressed in a rabbit reticulocyte lysate system, 40 kd polypeptides were produced that were recognized by polyclonal antisera directed against *Torpedo* α subunits. It is interesting to note that the two *Xenopus* α subunits are as similar to each other as either is similar to *Torpedo* α (~80% identity).

Analysis of the α1a (200) and α1b (150) transcripts indicates that both are expressed throughout early development, however, the ratio of α1a to α1b varies at different stages (Hartman and Claudio, manuscript submitted). It has been well documented in *Xenopus* that AChR-channels with at least two different conductances are expressed (23). *Xenopus* AChRs in newly differentiated muscle cells (up to about stage 25) exhibit 40 pS channels. At about stage 25, a second class of channel appears with a larger conductance of 60 pS which becomes the predominant channel type at later developmental stages. Although it is premature to speculate before developmental and functional studies are completed, it is none the less intriguing to postulate that the 40 pS channel may contain an α1a subunit and the 60 pS channel may contain an α1b subunit.

Because of the isolation of two 'apparently' novel AChR subunits (ϵ and a second muscle α), some caution should be exercised before identifying and naming the subunits. If there is a developmental switch between two subunits, it may be less confusing to refer to the subunits as fetal and adult rather than assigning them new Greek symbols. In contrast, if distinct subunit types are expressed at the same time in development in the same or different tissues, they might be named such that an altered structural or functional property will be reflected (e.g. channel conductance). As will be discussed in Section 10, multiple neuronal α and non-α subunit cDNAs have been cloned and sequenced. An effort will soon have to be made by all in the field to define, distinguish and standardize the nomenclature used to define muscle and neuronal AChR subunits.

7. Expression systems

Perhaps the best approach to correlating structure with function is to clone and sequence the gene of a protein with a known altered phenotype and compare it with the wild-type gene sequence. Without such naturally occurring mutations, an alternative approach is to make specific or random mutations in genes or cDNAs, express the altered DNA, and determine how the structure or function of the protein has been altered. In this later approach, a number of experimental systems have been used to express the altered DNA.

7.1 *In-vitro* translation systems

AChR subunits are not properly processed and assembled into functional complexes in *in-vitro* translation systems such as the rabbit reticulocyte lysate or wheat germ systems, and so these expression systems are not useful for studying assembly or structure/function relationships. Cell-free translation systems are, however, capable of producing AChR polypeptides in a very short time which has several useful applications in the study of AChRs. As discussed throughout this chapter, subunits from different species, novel subunits and neuronal AChR subunits have been isolated using *Torpedo* cDNAs as the initial probes. When crossing species boundaries or isolating novel clones, low stringency hybridization conditions must be employed which means that false positive clones will probably also be isolated. One method for determining if an uncharacterized clone is full-length and in the correct orientation, or if it is recognized by either species-specific or species-cross-reacting antisera, is first to prepare RNA *in vitro* using the clone as template. The RNA can then be put into one of the above cell-free translation systems to determine whether it will direct the synthesis of an appropriately sized polypeptide that can be immunoprecipitated with antibodies against receptor subunits. Once a clone has been identified, further analysis by sequencing and introduction into an expression system that produces functional molecules is then necessary. New lambda phage cDNA cloning vectors which also contain bacterial plasmid sequences (phagemids) have recently been designed (220–222) which allow high-efficiency directional cDNA cloning and selective amplification of either sense or antisense cRNA sequences (221). Using these cloning systems, one can very quickly obtain a recombinant plasmid which can be used to prepare mRNA *in vitro*. This RNA could be used in cell-free translation systems or for injection into oocytes (Section 7.3), as well as for various hybridization procedures.

7.2 *Escherichia coli*

The α subunit of the AChR contains the binding site for snake neurotoxins. Several studies using proteolytic fragments of this subunit or synthetic peptides corresponding to different regions have defined this important functional domain (reviewed in ref. 223). In a recent approach, expression of fragments of the α polypeptide in *E.coli* as fusion proteins with the β-galactosidase or *trpE* gene products has allowed the production of large quantities of material for analysis in binding studies by toxin or antibodies (224,225). Thus, the production of fusion proteins in *E.coli* is a method that can be employed effectively for studies in which large quantities of fragments of the different subunit polypeptides are desired.

7.3 *Xenopus* oocytes

The *X.laevis* oocyte system has been used very successfully to express not only properly assembled and fully functional AChR channels, but also many other channels, receptors, enzymes and proteins. The best method for achieving high levels of expression (136,152,226) is to prepare RNA *in vitro* from cloned DNA downstream of a bacteriophage promoter such as SP6 or T7 (227). Depending on the vector construction, the purity of the RNA, the temperature at which oocytes are incubated with RNA and the season of the year, AChR complexes can be expressed from about 2 days to 2 weeks. One of the major advantages of this system is the rapidity with which expression can be assayed following injection. In addition, the system lends itself to several experimental manipulations. The oocytes can be voltage-clamped and whole-cell recordings taken without much difficulty. An ingenious technique has been developed by Methfessel *et al.* (228) to allow gigaohm seals, required for single-channel recordings, to be formed between the pipette tip and the plasma membrane. The follicle cell layer is first removed with collagenase treatment and then, just prior to making recordings, the injected oocytes are placed in a hypertonic salt solution causing them to shrink away from the vitellin membrane which can be mechanically removed with forceps. The resultant 'clean' but very fragile oocytes can then be used in single-channel recording experiments.

The oocyte system, therefore, is an excellent transient expression system which allows expression of fully functional AChRs and lends itself to several types of experimental manipulations. The short delay between injection of mRNA and expression also makes it an excellent system in which to express mutant AChRs. Because many mutations will result in non-functional molecules, it is highly desirable to have a system for rapidly screening potentially interesting mutations before undertaking a thorough analysis of the mutant channels. The ability to record from single channels has significantly increased the amount of information that can be obtained from expressing both wild-type and mutant channels in this system. On the other hand, there are some general disadvantages of transient expression systems which apply equally to the oocyte system. For example, expression must be established every time an experiment is performed. Every cell is not identical to every other cell, and certain types of experiments may require that the AChRs be continuously expressed. A specific disadvantage of the oocyte system is that high-level expression is seasonal and is generally not achieved with oocytes obtained in the summer months. Furthermore, for many types of pharmacological and biochemical experiments, very large quantities of AChRs are required and these cannot be obtained in oocytes which must be individually injected.

7.4 Yeast

Yeast offers an expression system in which large quantities of material are often very easy to obtain (229,230). Foreign proteins have been expressed in yeast in a stable or a transient fashion after introduction of either chromosomal genes or cDNAs on appropriate plasmids. Hess's laboratory has demonstrated that the *Torpedo* α (231), γ and δ (232,233) subunits can be expressed in a transient fashion in yeast and that each subunit is expressed on the cell surface but the orientation of the subunits is incorrect. Analysis of the expressed gene products by SDS–PAGE revealed that each of the subunits migrated with the proper molecular weight and that low affinity BuTx binding by the α subunit (10^{-7} M) was achieved (233). K.U.Jansen *et al.* (234) have recently achieved expression of the *Torpedo* β subunit in yeast. They have established several yeast strains which stably express all four subunits in the same cell by mating strains containing individual subunits. Similar expression of all four *Torpedo* AChR subunits (G.Yellen, personal communication) and mouse BC$_3$H-1 AChR subunits (C.Kung, personal communication) in yeast has also been achieved. If the subunits can assemble into functional cell surface receptors then this would become a very powerful system for the analysis of AChR structure and function.

If the yeast system can be employed, then the tremendous genetic advantages of this system can also be exploited. Yeasts are ideally suited for analysis of various post-translational processes and transport pathways because of the availability of large numbers of strains with well-characterized mutations at specific points in these different pathways (235–237). Yeast cells are also well known for their ability to produce large quantities of foreign proteins (230). If functional AChR complexes can be produced in yeast, then this might become a method for producing enough pure AChR for subsequent structural studies. Because functional AChRs have not yet been produced in yeast, it is too soon to know how valuable this system will become for expression of the highly processed and complex AChR molecule.

7.5 Stable expression in cultured cell lines

To date, the only expression system other than oocytes to produce correctly assembled and functional AChRs is stable expression in cultured cell lines (238). In this system, the four *Torpedo* AChR subunit cDNAs were engineered separately into mammalian expression vectors and then co-transfected with a selectable marker gene (thymidine kinase, tk) on a separate plasmid into mouse fibroblast L cells deficient in the production of a functional tk enzyme (Ltk$^-$ cells). Transfected cells were then put into selective medium and tk$^+$ transformants were isolated and grown into stable cell lines. A thorough analysis of one of these cell lines has demonstrated that functional cell surface AChRs are produced with about 40 000 BuTx-binding AChRs being expressed on the surface of each cell. These AChRs are properly assembled into $\alpha_2\beta\gamma\delta$ pentamers (239) and they display proper physiological and pharmacological properties. These properties include similar desensitization kinetics, mean channel open time, single-channel conductance, the same rank order of affinities for agonists and antagonists, a dissociated constant for BuTx of 7.8×10^{-11} M, and reversible antagonists exhibit different affinities for the two binding sites on the AChR (238,240; S.M.Sine and T.Claudio, manuscript in preparation). One difference observed in toxin-displacement experiments was that the equilibrium binding of ACh and carbamylcholine had larger dissociation constants and Hill coefficients than had been reported in *Torpedo* membrane fragments. If one interprets these data in terms of an allosteric model for desensitization of the AChR, the weaker binding observed in the AChR-expressing fibroblasts could represent a shift of the allosteric equilibrium away from the desensitized state. Whereas in *Torpedo* electroplaque membrane fragment preparations 10–30% of the AChRs are in the desensitized state in the absence of agonist (241,242), the binding parameters obtained for AChR–fibroblasts would predict that only about 0.01% of the receptors are desensitized in the absence of agonist. Indirect evidence suggests that the values obtained for *Torpedo* AChRs in mouse fibroblasts probably represent the proper extent of desensitization for this receptor. This conclusion was reached because it was possible to perform, for the first time, competition binding studies between toxin and agonists on *Torpedo* AChRs in an intact membrane environment (238). Such studies cannot be performed with intact electrocytes because of the nature of the tissue, or with oocytes because sufficient quantities of material cannot be obtained.

There are several other advantages of the stable expression system over the transient oocyte system in addition to the ability of performing pharmacological and biochemical experiments. This system readily lends itself to investigating the biosynthesis, assembly (243,244; H.L.Paulson and T.Claudio, manuscript submitted) and modulation (169) of the AChR. Interactions between receptors and nerve, nerve trophic factors and cytoskeletal components can also be investigated. Fibroblasts continuously expressing AChRs on their cell surface can be co-cultured with nerves to investigate events of synaptogenesis such as nerve-induced AChR clustering and nerve–muscle adhesion (239,245).

Perhaps the biggest disadvantage of the stable expression system is that it takes up to 2 months to establish each cell line. This alone suggests that it would not be practical to combine site-directed mutagenesis experiments with stable expression, as many mutations result in non-functional AChRs. The problem can be simplified by first establishing and characterizing a cell line that expresses just three of the four subunits. The characterized cell line then serves as the recipient cell for introducing the fourth wild-type or mutant gene. Using such an approach, we have established, by co-transfection, a fibroblast cell line that stably expresses *Torpedo* β, γ and δ subunits. The α subunit was then stably introduced into this line using a packaged recombinant retroviral vector carrying the α cDNA (151). Cell lines have been isolated that express functional AChRs on the cell surface that can be analyzed at the single channel level. Although this approach greatly reduces the amount of work involved in characterizing the lines, it is still best to first test a mutation for functionality by expressing it in a transient system. Once an interesting mutation has been identified, introduction into an established stable cell line using viral infection, transfection, electroporation or lipofection with a second selectable marker will both reduce the amount of characterization needed and will provide a more uniform background from which to analyze the effects of the mutation.

8. Determining the function of individual AChR subunits

The AChR subunits are impossible to dissociate without the use of strong denaturing agents which then render the AChR non-functional. Thus, in order to study novel combinations of subunits, the appropriate cloned sequences have been expressed in *Xenopus* oocytes or mammalian cell lines. Two approaches have been used in attempts to determine the function of individual subunits: (i) expression of fewer than the full complement of subunits; (ii) expression of hybrid (also chimeric) AChRs composed of subunits from species whose AChRs display different properties.

8.1 Different combinations of AChR subunits

After the four *Torpedo* AChR subunit cDNAs were isolated and it was shown that the RNA transcribed from these clones expressed functional AChRs when injected into oocytes, a series of experiments were conducted to see which combinations of subunits might produce assembled and functional AChRs (246). Assembly was tested by solubilizing the oocytes and performing [^{125}I]BuTx-binding assays. Function was tested by measuring the whole-cell current in response to bath-applied ACh. The results demonstrated that all four subunits were required in order to obtain high level expression of assembled and functional receptor-channels. The combination of $\alpha\beta\gamma$ did, however, give 5% of normal activity, and the combination of $\alpha\beta\delta$ gave 3% of normal activity. In a later, more thorough study in which all combinations of subunits were presented, the additional assay of cell surface [^{125}I]BuTx binding was performed (247). The data presented were in general agreement with the above findings giving 10 and 0.3% of the normal response to 1 μM ACh for $\alpha\beta\gamma$ and $\alpha\beta\delta$, respectively. In addition, although no functional AChRs were established with any other combination of subunits (with the possible exception of $\alpha\gamma\delta$ in which 0.08% of the wild-type current was obtained if 10 μM ACh was applied instead of 1 μM ACh), some surface toxin-binding activity was detected with all combinations of subunits that contained an α subunit (including α alone, 0.7%). Although toxin binding to isolated α subunits has been widely reported (151,224,225,231,248,249), the only other report of cell surface expression of individual α subunits has been in the yeast system (see Section 7.4) where the subunits apparently do not integrate correctly into the plasma membrane and the toxin-binding function is not reconstituted (231).

The results of [^{125}I]BuTx binding to oocyte extracts injected with all combinations of α-containing subunits were very intriguing in terms of trying to establish which subunits might induce a proper toxin-binding conformation in the α subunit. The results were as follows: $\alpha\beta$, 0.9% of the amount of toxin binding detected with all four subunits; $\alpha\gamma$, 26%; $\alpha\delta$, 8%; $\alpha\beta\gamma$, 69%; $\alpha\beta\delta$, 10%; $\alpha\gamma\delta$, 33%. It would appear that γ is the most effective subunit at promoting the toxin-binding conformation in the α subunit. The δ subunit is about one-third as effective as γ but because the γ and δ subunits are the most similar subunit pair (49% identical), one interpretation of these results is that the only reason δ works as well as it does is because it looks like γ. Some possible implications of this later interpretation and how it might be used to try to determine the configuration of subunits around the channel will be discussed in Section 8.3.

Because of the difficulty in expressing AChRs without all four subunits, not much information has been gained about individual subunit contributions to receptor functions with the exception of δ. An AChR composed of $\alpha\beta\gamma$ (δ-less) has 69% of wild-type ($\alpha\beta\gamma\delta$) ability to bind [^{125}I]BuTx. About a third of these receptors are expressed on the surface (36%), and the surface receptors respond to ACh with 10% of the current response of wild-type AChR (247,250). The findings that surface receptor expression is 36% of normal but the current response is only 10% of normal would indicate that the smaller currents are not entirely due to a decreased number of surface receptors. In a study of wild-type and δ-less AChRs expressed in oocytes, δ-less surface receptors were shown to have identical affinities for d-tubocurarine, but the channels had a different current–voltage relationship (250). The implication of these studies is that δ may play a role in suppressing voltage sensitivity of the channel.

8.2 Hybrid and chimeric AChRs

AChR subunit cDNAs have been isolated from a number of species (*Tables 1* and *2*) making it possible to inject different combinations of subunit mRNAs from different species into oocytes to determine if functional AChRs can be produced and if altered properties can be detected. Because cDNA clones have been widely distributed for *Torpedo* (130,132,133,135,136) and the mouse muscle cell line BC_3H-1 (149,186, 196–199), several groups have reported data obtained with *Torpedo*–mouse hybrids (226,243,251–254). In addition, a functional chicken–*Torpedo*–mouse hybrid has also been reported (253).

8.2.1 Torpedo–mouse hybrids

For all of the hybrid studies involving cDNA clones RNAs were synthesized *in vitro* using the SP6 polymerase system (227). Different combinations of subunit RNAs were then injected into *X.laevis* oocytes. It was first determined that mouse δ plus *Torpedo* αβγ subunits could assemble into cell surface functional receptors and that these receptors showed a 3- to 4-fold greater response to ACh than all-*Torpedo* AChRs (226). An extensive study of the 16 possible combinations of mouse and *Torpedo* subunits revealed that the mouse β subunit induced the greatest voltage sensitivity in channels (252). In Section 8.1, however, data were presented indicating that low voltage-sensitivity was due to the δ subunit (250). These seemingly different results could be reconciled as follows. Voltage sensitivity can be due to altered channel conductance, opening rate, or closing rate (mean open time). Using single-channel analysis, Yu *et al*. (254) determined that the closing rate made the greater contribution to voltage sensitivity in the mouse β-hybrid AChRs. Further analysis of the δ-less mutant receptors may reveal that δ exerts its effects on channel conductance or opening rate and not on closing rate.

8.2.2 Torpedo–calf hybrids and chimeras

A series of studies were conducted in which the four AChR subunits from calf, the four subunits from *Torpedo*, or different combinations of calf and *Torpedo* subunits were expressed in oocytes after microinjection of RNA transcribed *in vitro* (207,255). When the channel properties of all-calf and all-*Torpedo* were investigated in oocytes using normal physiological concentrations, no differences in conductance were observed, but the channel open times differed by an order of magnitude: 7.6 ms for calf and 0.6 ms for *Torpedo* (255). The results of mixing various combinations of calf and *Torpedo* subunits demonstrated that AChRs composed of *Torpedo* αβγ plus calf δ had channel open times (8.6 ms) similar to those of all-calf AChRs (255). The implication from these studies was that the δ subunit was responsible for determining the open time of the channel. These results were very exciting because it was assumed by many investigators that such channel properties would probably be under the control of the α subunit since the α subunit was the location of the agonist and antagonist binding sites. The results of the reciprocal experiment (calf αβγ plus *Torpedo* δ) were not reported, thus a stronger case for the role of δ in gating could not be made.

In a later study by the same authors (207), a very elegant and laborious set of experiments was reported in which a series of 11 chimeric δ subunits had been constructed between *Torpedo* and calf sequences and expressed with *Torpedo* αβγ subunits. If the channels formed had been analyzed in terms of their open times then the role of δ in gating might have been more clearly defined; instead, channel conductances were assayed. The single-channel conductances were measured using pipettes filled with solutions devoid of divalent cations (Ca^{2+}). In the absence of external Ca^{2+}, all-calf, all-*Torpedo* and the chimeric AChRs fell nicely into two classes of conductance: calf-like (64–72 pS) and *Torpedo*-like (85–88 pS). If physiological concentrations of Ca^{2+} (~2 mM) were present, no differences in conductance were detected between all-calf and all-*Torpedo* AChRs, nor were differences detected among the different chimeras. One δ chimera composed of approximately 500 amino acids of *Torpedo* and only 29 amino acids of calf produced a conductance of 67 pS (calf-like) whereas a chimera composed of approximately 500 amino acids of calf and only 36 amino acids of *Torpedo* produced a conductance of 88 pS (*Torpedo*-like). The sequences responsible for producing the conductance changes consisted of the M2 domain plus the bend region between M2 and M3. Examination of these sequences in *Torpedo* and calf indicated that the M2 domains of these two species were very similar. The major differences between species in the bend region between M2 and M3 were the presence of two positively charged residues in calf compared with one positively and one negatively charged residue in *Torpedo*. The observations that a small amount of negative charge is probably present near the opening of the channel (203,204) and the results of this study are consistent with the hypothesis that the M2–M3 bend region is located near the opening of the channel and that the negatively charged residues are attracting permeant cations toward the mouth of the channel (207).

In yet another study by Imoto et al. (208), a large number of point mutations were made in each of the four *Torpedo* subunits. Both outward and inward currents could be affectd by changing the charge in primarily three regions of negatively charged and glutamine residues: two sites between M1 and M2 and one site between M2 and M3. Although these and the above experiments had to be performed under non-physiological conditions in order to detect any differences in conductance, the results are, none the less, very intriguing. The combination of exhaustive site-specific mutagenesis with detailed electrophysiological analysis has allowed the Numa and Sakmann groups to alter channel conductance and gain some insight into the structure and function of this receptor-channel (207,208,255).

8.2.3 Expression of AChRs with ϵ subunits

As discussed in Section 6, a novel subunit, ϵ, was isolated from a calf cDNA library. When mRNAs were prepared *in vitro* from each of the five calf cDNAs (α, β, γ, δ, ϵ) and different combinations of four were microinjected into oocytes, it was determined that ϵ could substitute for γ but not for any other subunit. The conductance of channels composed of $\alpha\beta\gamma\delta$ was calf-like (40 pS) while the conductance of $\alpha\beta\epsilon\delta$ channels was cow-like (60 pS) (218). The authors concluded that the γ and ϵ subunits were responsible for the different conductances observed. However, data presented in the previous section (8.2.2) indicated that the region between M2 and M3 of the δ subunit was also important in determining the rate of ion transport through the channel (207). Is channel conductance determined by the γ or the δ subunit? I believe it is fair to say that, as yet, no structural correlates of channel properties have been unambiguously identified. It is also quite likely that simple explanations for these properties will not be forthcoming. If, for example, one transmembrane domain of each subunit contributes to the lining of the ion channel, then each subunit may make a significant contribution to the conductance of the AChR. Although altering one domain of one subunit might drastically alter channel properties thereby suggesting that subunit's contribution to a particular property, altering the homologous domains of the other subunits might similarly affect the channel.

8.2.4 Torpedo-endogenous mouse hybrids

Torpedo α subunits have been stably introduced into two mouse fibroblast, two mouse muscle and one rat muscle cell line (30,136,151,238,243, 256,257; H.L.Paulson and T.Claudio, manuscript submitted; see Section 7.5). The stable introduction of mouse α cDNA into quail fibroblasts has also recently been reported (258). One of the three muscle – *Torpedo* α cell lines has been analyzed for the expression of hybrid AChRs (243,244). The host cell used for these studies was the rat muscle L6 (259) cell line. When L6 – *Torpedo* α cells were allowed to differentiate and express endogenous rat AChRs, three types of AChRs could be detected using species-specific anti-α antibodies: (i) endogenous AChRs in which the two α subunits were both of rat origin; (ii) hybrid AChRs in which the two α subunits were of *Torpedo* origin; (iii) hybrid AChRs containing one rat and one *Torpedo* α subunit (243,244). One conclusion that could be made from these results was that the two α polypeptides in an AChR pentamer need not derive from the same polysome.

In none of the hybrid experiments conducted in oocytes has there been a competition between like-subunits from different species for assembly into AChR complexes. One of the interesting results from the L6–Torpedo α cell lines was that hybrid AChRs were expressed in the presence of the full complement of four rat subunits. In addition, hybrids were expressed in an environment that was presumably optimal for expression of rat AChRs. These results imply that the tertiary structures of *Torpedo* and rat α subunits are very similar and may be more similar than is indicated by the 80% primary amino acid sequence identity observed between them (*Table 2*). It is also possible that the region of α in *Torpedo* and rat that makes contact with the other subunits is identical (see also Sections 8.3 and 8.4).

8.3 Configuration of subunits around the channel

The AChR has a very unusual subunit composition and stoichiometry with five subunits in the stoichiometry, 2:1:1:1. Intriguing pharmacological observations have been made concerning the non-equivalence of the two α subunits. The affinity label, MBTA, reacts at very different rates with the two ACh binding sites and competitive antagonists bind with different affinities, indicating that the two α subunits are functionally different (1,260). Several possibilities for the different α behaviors could be postulated: (i) the primary sequences of the two α subunits could be slightly different; (ii) there could be different post-translational modifications of the two α subunits; (iii) the α subunits could be identical both in primary sequence and post-translational modifications but the subunits display different ligand-binding properties because each α subunit is flanked by a unique pair of subunits.

The first possibility appears unlikely because only a single gene has been found for α (261,262) indicating that the primary amino acid sequences are identical [the possibility of two α subunits in *Xenopus* (203) is discussed in Section 6.2]. There have been several reports suggesting that structural differences exist between fast (junctional or adult) and slow (extrajunctional or embryonic) channels (109–113,263). However, although one report suggested a possible glycosylation difference between the two α subunits of a single AChR complex (139), other studies indicated that there were no such differences (264). A very likely explanation for the pharmacological differences observed in α subunits of the same molecule is that the different properties are induced by the local environment, namely interaction with those subunits which border each α subunit. The arrangement of subunits around the central pore is a question of great interest because of its relevance to subunit–subunit interactions, subunit assembly, and channel structure.

8.3.1 *Evidence indicating that γ lies between the two α subunits*

Several studies have indicated that the two α subunits do not lie next to each other and that the δ subunit is not positioned between them. The location of the α subunits relative to each other was determined by using a complex of biotinylated cobra (*Naja naja siamensis*) toxin (which binds to α subunits) and avidin (which has a molecular weight of 67 kd and binds to biotin) as an electron microscopic marker for α chains in negatively stained AChR monomers. By measuring the angle between the avidin molecules, it was determined that the two α subunits were not adjacent. A similar approach was taken in order to determine the location of α subunits relative to the δ subunit except that the binding was to AChR dimers (267,268). In that study, the position of avidin molecules relative to the region of closest approach of two receptor molecules (presumably the disulfide linkage between δ subunits) was used to determine the location of δ subunits relative to α subunits. From these studies it was concluded that α subunits were not located next to each other and that δ subunits were not located between the α subunits (267,268). Similar conclusions were reached using image reconstructions of negatively stained AChRs in membranes (204) and the change in stain distribution due to toxin binding (269).

In order to determine the location of β subunits, dimers were artificially produced between β subunits such that β–β cross-linked dimers could be used in an electron microscopic study with biotinylated toxin and avidin in a fashion similar to that used for the naturally occurring δ–δ cross-linked dimers. The angles measured between avidin molecules and the β–β cross-link indicated that β was not located between the α subunits (270). It was therefore assumed that γ must be located between the α subunits.

8.3.2 Evidence indicating that β lies between the two α subunits

Receptor-specific ligands modify the α-neurotoxin cross-linking pattern in a way that suggests that β, rather than γ, may lie between the two α subunits (271). Other interpretations have also supported this later view (133,272). In another approach, electron image analyses of two-dimensionally ordered arrays of AChRs formed from receptor-rich membranes having a tubular morphology were used to determine the arrangement of subunits around the pore (273). BuTx or the Fab fragment of a monoclonal antibody directed against an extracellular epitope of the α subunit [mAb35 (274)] was used in order to mark the α subunit; the β subunit was tagged with the Fab fragment of a monoclonal directed against its cytoplasmic domain [mAb111 (275)]; and the δ subunit was marked with wheat germ agglutinin which binds specifically to *N*-acetylglucosamine moieties located exclusively on the δ subunit (144). Tubular membranes containing AChR were prepared, specific subunit tags were incubated with the tubes, electron micrographs were taken, and images were analyzed by Fourier transformation and statistical difference mapping. This analysis indicated that the β subunit was located between the α subunits.

Although this very elegant study has provided the strongest evidence for the arrangement of the subunits, a few doubts still remain. This study, like all others, came to conclusions about the order of subunits without determining the location of at least one of the subunits, in this case the γ subunit. Also, there is concern about antibody cross-reactivity and specificity. For example, mAb111 which is directed against β appears to cross-react with α. The question is: what epitopes are being recognized by mAb111, which is directed against a cytoplasmic determinant, when it is incubated with AChR tubes? The conclusions of this study would be unrefutable if other β antibodies and some γ antibodies could be found that would react with the AChR tube preparations described in this study.

8.3.3 Subunit deletion experiments

The results of a study by Kurosake et al. (247) are most intriguing especially when trying to interpret them in terms of their relevance to the subunit arrangement problem. In this study, all combinations of the four *Torpedo* subunit RNAs were injected into oocytes and the ability to bind BuTx was monitored. The results obtained after solubilizing the oocytes and assaying for [^{125}I]BuTx (100% toxin binding being defined as that amount obtained after injection with αβγδ) were as follows: α, 1%; αβ, 1%; αγ, 26%; αδ, 8%; αβγ, 69%; αβδ, 10%; αγδ, 33%. The results indicate that the γ subunit is the best subunit for inducing a toxin-binding conformation in the α subunit (26%) and that the β subunit is incapable of inducing this conformation (1%). The δ subunit has an intermediate value (8%) which could be a reflection of its high degree of identity with γ (49%) thus allowing them to more readily substitute for one another. Thus, one interpretation of these results is that the γ subunit is capable of inducing the toxin-binding conformation but the β subunit is not.

Other data relevant to this interpretation are that the most difficult hybrid AChRs to establish are those in which one tries to substitute the β or γ subunit. This could indicate that the most specific subunit interaction is dictated by the β or γ subunit. Since the easiest subunit to substitute is δ, and the γ and δ subunits are so similar, one might expect that γ would substitute as readily as δ, but it clearly does not. If one proposes a model of the subunit order about the channel with a minimal number of specific subunit–subunit contacts (274; see *Figure 2*), then non-adjacent subunits in which one of the subunit interaction sites is in common would be more likely to substitute for one another than if both subunit contact points were different. As drawn in *Figure 2*, if γ were adjacent to δ, no common contact points between these two subunits would result. In contrast, if γ were located between the two α subunits, γ and δ would have one contact region in common. This line of reasoning would favor the γ subunit being located between the two α subunits.

8.4 Speculations on the order of subunit assembly

It is not known whether the subunits of the AChR assemble with a specific order or if subunit assembly into proper $\alpha_2\beta\gamma\delta$ pentamers is a random

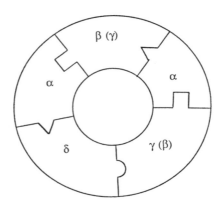

Figure 2. Configuration of subunits around the channel. The model is the same one presented by Karlin in ref. 276. A minimal subunit interaction model is presented in which the specific contact points between adjacent subunits are denoted by the half-square, -diamond, and -circle symbols. Alternative subunit arrangements are shown in parentheses.

process. It has been shown in the mouse muscle cell line BC_3H-1 that unassembled subunits migrate on sucrose gradients with a sedimentation coefficient of about 5S (275) suggesting a molecular weight of a dimer or a trimer. No associations between different subunits indicative of possible assembly intermediates, short of the pentameric complex, have been identified. One of the difficulties of addressing questions of assembly in these muscle cell lines is that assembly cannot be synchronized, making it virtually impossible to isolate large quantities of any intermediate complexes. Large quantities of possible assembly intermediates can be artificially produced using the stable expression system described in Section 7.5.

Analysis of different combinations of subunits stably expressed in cell lines will establish if certain subunit interactions are more favorable than others. One interpretation of such results is that they reflect assembly intermediates. The isolation of actual intermediates is more difficult. We may be able to isolate true intermediates, however, using our *Torpedo* AChR-expressing fibroblast cell lines. Expression of functional *Torpedo* AChRs in fibroblasts is acutely sensitive to temperature (238). Although subunits are synthesized at normal mammalian growth temperatures (37°C), complexes are not formed unless the temperature is reduced by approximately 10°C (244). Using this temperature-sensitive system for AChR expression, we may be able to synchronize the assembly process and isolate intermediates by growing the cells at 37°C and then shifting to 27°C.

It is hoped that the above types of experiments will define specific subunit–subunit interactions and address the question of the order of subunit assembly. In the meantime, I will attempt to predict the answers to these questions using published observations. It should be duly noted that this discussion is riddled with assumptions and is highly speculative. It is, however, an attempt to combine data collected on expression of *Torpedo* – mouse and *Torpedo* – bovine hybrid AChRs with data collected on expression of different combinations of *Torpedo* subunits. It will put forward the notion that subunits are assembled in a specific order and speculate on what that order might be.

The first assumption made is that the more toxin binding that is observed, the more closely the α subunit approximates its normal assembled ($\alpha_2\beta\gamma\delta$) conformation. The experiments described in Section 8.3.3 concerning various combinations of *Torpedo* subunits expressed in oocytes (247), clearly demonstrate that certain subunits can effectively induce a BuTx-binding conformation in the α subunit, suggesting that subunit – subunit interactions exist and possibly indicating that there are specific subunit interactions. If there are such specific subunit – subunit interactions, then it is possible that there is an order to the assembly process. One subunit might induce a conformation in a second subunit resulting in conformational changes which allow the third, fourth, and then the fifth subunit to assemble. Because of the striking effect γ has on the ability of α to bind BuTx, I propose that the first interaction is between α and γ subunits. The results from two studies in which hybrid AChRs composed entirely of *Torpedo* subunits except for one bovine subunit (255), or entirely *Torpedo* subunits except for one mouse subunit (252) were consistent with each other in that α and δ subunits could be substituted far more readily than either β or γ. The ease with which a subunit from one species can substitute for its homologous subunit from another species could be due to either (i) a high degree of subunit similarity between the two species; (ii) the lack of a significant structural or functional role for the subunit in the assembled complex enabling it to conform to an altered configuration; or (iii) the assembly of the subunit may occur earlier than others, thus restricting the subsequent assembly of other subunits into the complex.

The similarity between *Torpedo* and bovine or murine subunits is 80, 58, 54 and 59% for α, β, γ and δ, respectively. Whereas the first above explanation (i) could account for why α can be substituted far more readily than β or δ, it cannot explain why δ can be substituted far much more readily than β (5.4 fmol [^{125}I]BuTx binding sites on the surface/oocyte versus 0.1 fmol/oocyte for *Torpedo* – bovine hybrids) since both are equally similar to their homologs. There is the possibility, of course, that there are regions of identity at specific subunit – subunit interaction sites that would not be detected in the analysis of the entire subunit. No information is yet available that can address whether one subunit is more malleable than another, and thus nothing can be said concerning explanation (ii) postulated above. Concerning the third above explanation (iii), if there is an order of assembly, then it may be that β is so difficult to substitute because it is the last to assemble and therefore the most physically constrained. My proposed order of assembly is, therefore, α, γ, δ and finally β. As will be discussed briefly in Section 9.2, my laboratory has found that the half-life of the β subunits is much shorter than the half-lives of the other three subunits when stably expressed in cultured cells (151,256). This could indicate that β plays a regulatory role in assembly and, if so, then the regulation could be at the final step in the assembly process, that is, β is the last subunit to be added to the complex.

9. Isolation of chromosomal genes for nicotinic AChRs

In addition to the large number of AChR cDNAs, the chromosomal genes for several AChR subunit genes have also been isolated. These include: human α (193,276) and γ (147); *Torpedo* δ (277); chicken α (278–281), γ and δ (146); rat γ and ϵ (219); and mouse γ (281) and δ (282,283). The *Torpedo* cDNAs were instrumental in obtaining these various genes in that they were used either to screen genomic libraries directly or to obtain a cDNA (such as calf) which in turn was used to screen the genomic library (such as human).

9.1 Genomic organization

Genomic blots using cDNA probes from different species all indicate that there is a single gene for each of the four subunits. It has been shown that the chicken γ and δ genes (146) and the human γ gene (147) each contain 12 exons. In contrast, the human (193,276) and chicken (278) α genes consist of only nine exons each. The human and chicken α genes have identical patterns of organization and the human γ, chicken γ and chicken δ genes also have the same exon–intron arrangements. There is no apparent homology between the intervening sequences of like-subunits from different species nor are the lengths of the introns conserved. The chicken γ and δ genes are closely linked, being separated by only 740 bp, with the δ gene being located upstream of γ (146). In human, these two genes also appear to be closely linked (147), and they co-segregate on the same chromosome in mouse (284). The chicken α exons extend over about 4.3 kb (279), human α over about 16.7 kb (193), chicken γ over about 4 kb (146), human γ over about 6.2 kb (147), chicken δ over about 3.3 kb (146), and *Torpedo* δ may extend over about 20 kb (277). Some of the protein segments encoded by the different exons tend to fall into structural or functional domains. For example, the signal sequence, M1 and M4 domains are each encoded by separate exons in the α, γ and δ genes. M2 and M3, on the other hand, are both encoded within the same exon in α but in two separate exons in γ and δ. Similarly, the region between M3 and M4 is encoded by one exon in α and two exons in γ and δ. The chromosomal locations of the subunits have been identified as chromosome 17 for mouse α, chromosome 11 for mouse β, and chromosome 1 for the mouse γ and δ genes (284).

9.2 Transcription of AChR genes

Almost 20 years ago it was demonstrated by Fambrough (285) that inhibition of RNA or protein synthesis prevented the development of ACh supersensitivity (see Section 2.4; refs 16–21) in organ cultures of denervated rat diaphragm, indicating that gene activity was required for the increased AChR expression observed. Now that specific probes for each of the subunits are available, several groups have been able to determine precisely the levels of expression of different subunit RNAs during innervation, denervation and reinnervation. Denervation leads to increased levels of α subunit mRNA in mouse leg muscle (286,287), chicken leg muscle (288–291), rat diaphragm (287) and *Xenopus* leg muscle (150); increased levels of γ subunit mRNA in chicken leg muscle (290); increased levels of δ subunit mRNA in mouse leg muscle (291) and chicken leg muscle (290); and increased levels of ϵ mRNA in rat diaphragm (219). AChR subunit mRNA levels have also been measured *in vitro* on cultured chick myotubes [α expression (291); α, γ and δ expression (290)] and murine myogenic cell lines [α and δ expression (292)] using various treatments which inhibit or enhance AChR expression. The results of the *in-vitro* experiments were in general agreement with those obtained from denervated muscle. Using denervated chicken leg muscle, Shieh et al. (290) demonstrated that α, γ and δ mRNA levels were each increased (no information on β was provided), by 112-, 42- and 24-fold, respectively, while the level of expression of AChR increased approximately 150-fold. A more thorough analysis of the α RNA indicated that there was an accelerated rate of transcription (as opposed to an altered turnover rate) which could account for the increased RNA content and the increased synthesis of AChR (289).

The disparity between different subunit mRNA levels, and these levels and AChR levels, suggests that AChR expression is regulated by mechanisms other than a simple coordinated increase in each of the subunit mRNAs. Several regulatory mechanisms could be in effect at the levels of transcription, RNA processing, translation, protein stability and assembly. Very little information has been presented on the regulation of expression of the β subunit. In fact, only one recent report indicates that β transcription increases in rat muscle after denervation; however, the increase in β transcription is much less pronounced than with α, γ or δ (293). The current lack of quantitative data on this subunit leaves open the possibility that it could be regulated differently from the other subunits. Interestingly, in my laboratory we have observed that the β subunit has a much shorter protein half-life than any of the other subunits suggesting a possible regulatory role for this subunit. Protein half-life studies of the *Torpedo* subunits each stably expressed in a different fibroblast cell line indicated that the half-lives at 37°C for α, γ and δ were each about 43 min, whereas for β it was only about 12 min. At 28°C (a temperature at which functional *Torpedo* AChRs are expressed in mammalian cells), the half-lives for α, γ and δ were about 70 min, and for β about 50 min (151,256).

A system that had traditionally lent itself to studies of embryogenesis and development is that of the African clawed toad, *X.laevis*. A wealth of information has also been accumulated on *Xenopus* concerning muscle differentiation and synapse formation (294) as well as detailed electrophysiological analysis of the AChRs (295; reviewed in ref. 23). Baldwin et al. (150) were able to isolate *X.laevis* AChR α, γ and δ (but not β) subunit clones from stage 17 and stage 22–24 embryo cDNA libraries using the four *Torpedo* subunit cDNAs (136) as probes. The mRNA levels of each of the three subunits have been measured during different stages of development by using radiolabeled antisense probes of the clones in RNase protection assays. The onset of transcription of each of these subunits was comparable to that of actin expression beginning at stage 17 (late gastrula). The amounts of AChR transcripts increased in parallel over the next 22 h up to stage 20, the initial stage of synapse formation. At stage 25, with the onset of motor activity, α, γ and δ transcripts each began to decrease. However, by stage 41 when tadpoles are freely swimming, α and δ transcripts decreased 4-fold, whereas γ mRNA decreased 55-fold per somite. These results demonstrate that coordinate expression is observed at some stages of development but apparently not at others, and suggest that different mechanisms of regulation may be called into play at different developmental times. The possibility of two α subunits in *Xenopus* was discussed in Section 6.2.

Recently, the 5' regulatory regions of several subunit genes have been isolated thereby enabling the identification of sequences responsible for tissue specificity, coordinate and developmental regulation of expression, and the effect of muscle activity and innervation on expression. Thus far it has been shown that the 5' flanking region of the chicken α gene (280), the mouse γ gene (281) and the mouse δ gene (282,283) can confer cell type specificity and developmental regulation.

10. Isolation of neuronal AChRs

Two strategies have been employed for isolating neuronal AChR genes and cDNAs. One approach was to prepare a library from the species of choice then screen it with *Torpedo* cDNAs in order to isolate the skeletal muscle AChR cDNAs. This cDNA was then used to screen for cDNAs for the neuronal AChR (148,279). The other approach was to screen directly for neuronal clones using *Torpedo* cDNAs (296). Thus far, four rat neuronal clones have been identified: $\alpha 2$ from rat brain (279,297,298); $\alpha 3$ from the rat pheochromacytoma cell line PC12 (148); $\alpha 4$ ($\alpha 4$-1 and $\alpha 4$-2) from rat brain (299); and non-α or $\beta 2$ from PC12 cells (300,301). Five chicken neuronal clones have been identified: $\alpha 2$, $\alpha 3$, $\gamma 2$ (279), $\alpha 4$ and non-α (302). In addition, two *Drosophila melanogaster* neuronal clones have also been identified: non-α or ARD (296) and α or ALS (303).

The neuronal clones were not as easily identified as the skeletal muscle clones because of sequence divergence (see *Tables 2* and *3*). Essentially, any neuronal clones which contain two adajcent cysteine residues analogous to *Torpedo* α Cys-192 and -193, located at or near the ACh binding site, are designated as α subunits. The non-α clones do not contain these cysteine residues and are thus referred to either as non-α or as β. It should be noted, however, that the designation β does not imply that it is particularly similar to skeletal muscle-like β subunits. Of the neuronal clones for which sequence information has been published, rat $\alpha 3$, rat $\alpha 4$, *Drosophila* non-α, chicken $\alpha 2$, $\alpha 3$, $\alpha 4$ and non-α each contain Cys-128, Cys-142 and the proline residue located in the middle of M1. Thus far, these three residues have been found in all AChR subunits of all species (see Section 5.2). One possible method for identifying neuronal clones would be with functional assays. As discussed in Sections 6.1 and 8.2.3, the calf ϵ subunit was identified as being γ-like because it was able to substitute for *Torpedo* γ subunit (*Torpedo* $\alpha\beta\delta$ plus calf ϵ) and form functional channels in oocytes (217). Unfortunately, the neuronal and muscle subunits apparently have diverged sufficiently as to prevent (in most instances) the formation of hybrid AChRs and so identifying the specific subunit type using this method has not been possible on a routine basis. Although all combinations were not tested, in one report (301), neuronal $\beta 2$ could substitute for mouse muscle BC_3H-1 β in functional assays suggesting that neuronal $\beta 2$ was in fact a β-like subunit. Because of the difficulty of expressing functional hybrid AChRs composed of neuronal and muscle subunits, another possible method for subunit identification might be at the level of genomic organization. As yet, no chromosomal β gene has been isolated. However, the muscle α genes have been shown to be encoded by nine exons whereas the γ and δ genes are each encoded by 12 exons. Analysis of the genomic organization of chicken neuronal $\alpha 2$, $\alpha 4$ and non-α genes as well as rat genomic $\alpha 2$ and $\alpha 3$ genes (302) revealed that each had identical structures consisting of six protein-encoding exons. These results would indicate that the neuronal AChR genes have similar genomic organizational motifs but they are distinct from the muscle-like genes.

The various neuronal clones have been shown to display neural specificity and to be expressed in different regions of the CNS (148,179, 279,304). Depolarizing responses to perfused ACh have been obtained in *Xenopus* oocytes after injection of mRNAs transcribed *in vitro* for the following rat clones: $\alpha 3 + \beta 2$, $\alpha 4 + \beta 2$, $\alpha 4$ alone (300); $\alpha 2 + \beta 2$ (297). Breer and colleagues have isolated a protein complex of molecular weight 250–300 kd from locust neuronal membranes, which migrates on SDS–polyacrylamide gels as a single polypeptide band of 65 kd (305). They were able to reconstitute a functional AChR into planar lipid bilayers with apparently just the 250 kd material which presumably was composed of only the 65 kd protein (306,307). The implication from these studies is that the insect neuronal AChR is composed of possibly four to five identical subunits. The results obtained from expression in *Xenopus* oocytes of cloned neuronal genes would also indicate that functional AChRs can be formed with just one or two different subunits. Now that neuronal AChR subunit genes have been isolated, a thorough electrophysiological, pharmacological and biochemical characterization is possible. The results of such studies will determine whether or not all of the pharmacological and physiological properties can be reconstituted with the clones isolated thus far, or if other subunits are required. If completely functional receptors can be reconstituted with the existing clones, then it will be very interesting to determine the stoichiometry of the subunits that form a functional AChR complex. This information should shed some light on the composition of the ancestral AChR and also on the evolutionary relationship between neuronal and skeletal muscle AChRs.

11. Other cloned channels

Several channels have now been cloned and sequenced. One of the most interesting findings to come from the sequence information is that channels with similar functions have similar structures. As discussed below, three ligand-gated channels have several structural features in common and three voltage-gated channels also have common structural features. The ligand-gated channels thus appear to belong to one large family or superfamily of channels and the voltange-gated channels appear to belong to another superfamily of channels.

11.1 Cloned ligand-gated receptor-channels

A brief introduction to ligand-gated receptor-channels can be found in Section 1.1. The three classes of neurotransmitter-gated ion channels which have been cloned thus far, include the ACh, the glycine (34), and the GABA receptors (35).

11.1.1 ACh receptors

As discussed in Section 5.2, muscle-like nicotinic AChRs isolated from the six different species studied are extensively similar (see *Tables 1* and *2*). Neuronal AChRs isolated from three different species also display similarities among themselves and with muscle-like AChRs (see *Table*

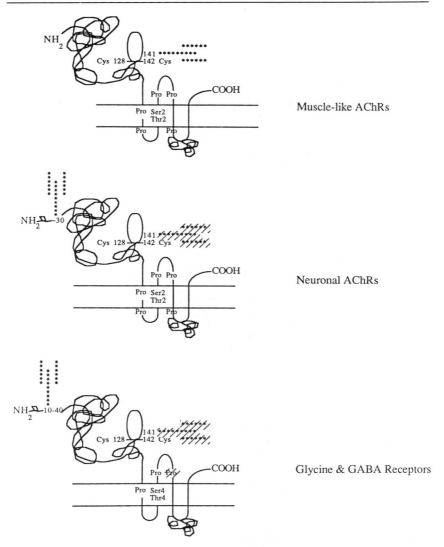

Figure 3. Models of the muscle-like AChRs, the neuronal AChRs and the glycine and GABA receptors. Only those features which are common to all of the members of each class of channels are shown in the figure. From hydropathy profiles, all of the channels contain four domains which are long enough and hydrophobic enough to span the lipid bilayer. Cysteine residues are present in all of the channels which are separated by 13 amino acids and are shown as linked residues numbered Cys-128 and -142. The series of stars represent potential sites of asparagine-linked glycosylation. Hatched marks (glycosylation sites or proline residues) indicate that some but not all of the members of this group of channels possess such a site or residue. The four to five serine and threonine residues present in M2 of the muscle-like and neuronal AChRs are denoted $Ser_2/Thr2$, and the seven to eight residues present in M2 of the GABA and glycine receptors are denoted Ser4/Thr4.

3). Common features for the 23 muscle-like AChR subunits isolated thus far include:
(i) the presence of four domains which are long enough and hydrophobic enough to span the lipid bilayer (M1 – M4, see *Figure 3*);
(ii) two cysteine residues separated by 13 amino acids and located in the extracellular domain preceding putative transmembrane domain M1 (Cys-128 and Cys-142 using the *Torpedo* AChR α subunit numbering system);
(iii) a potential asparagine-linked glycosylation site located at position 141;
(iv) a proline residue located in the middle of M1;
(v) a proline located just past M1;
(vi) two proline residues between M2 and M3;
(vii) a proline residue just past the M3 domain;
(viii) a large number of serine and threonine residues located in M2 (most subunits have 4 – 5 serine and threonine residues out of the 19 amino acids that comprise the M2 domain) (see *Figure 3*).

These features are also present in the 11 different neuronal AChR sequences that have been published thus far (both for α-like and non-α subunits) except for variations in the asparagine-linked glycosylation sites. There is at least one potential asparagine-linked glycosylation site in the *Drosophila* non-α (ARD) subunit (296), the *Drosophila* α (ALS) subunit (301), chicken neuronal $\alpha 2$ subunit and chicken neuronal $\alpha 4$ subunit (300), but this site is not located at position 141 but rather at about 30 residues from the amino terminus. Although rodent $\alpha 3$, $\alpha 4$ and non-α subunit ($\beta 2$) (298), as well as the chicken neuronal non-α subunit (303), each have a potential asparagine-linked glycosylation site at position 141 (the usual position for muscle-like subunits), they also have a site about 30 residues from the amino terminus (see *Figure 3*). In addition to the site at position 30, the chicken $\alpha 2$ and $\alpha 4$ also have a site at position 80.

11.1.2 Glycine receptors

The structure of the glycine receptor has not been precisely determined. There appear to be three polypeptides of molecular masses 48, 58 and 93 kd. The 48 and 58 kd subunits are glycosylated integral membrane proteins which appear to contain the ligand-binding sites and might form the ion channel. The 48 kd subunit contains the binding sites for strychnine and other antagonists whose binding can be inhibited by glycine and other agonists. Electrophysiological and pharmacological data indicate that more than one glycine molecule must bind in order to open the channel. The 93 kd polypeptide now appears to be a cytoplasmic peripheral membrane protein that can be separated from the other two subunits under reducing conditions (308). Its functional and structural roles are not known, and it is possible that it is not an integral component of the channel but it may perform roles similar to those proposed for the AChR-associated 43 kd protein. The 43 kd protein co-purifies under many conditions and co-localizes with the AChR. The function of this protein has not been determined but it is often proposed that it may help cluster or anchor the AChR at the neuromuscular junction (103). The uncertainty in the identity of glycine receptor subunits obviously makes it difficult to establish a stoichiometry. If the 93 kd polypeptide is not a part of the channel, then a possible stoichiometry of three 48 kd subunits and two 58 kd subunits would be consistent with the observed 250 kd complex (308).

The cDNA clone corresponding to the strychnine-binding (48 kd) subunit of the glycine receptor was isolated from a rat spinal cord cDNA library and sequenced (34). There are several striking similarities between the glycine 48 kd subunit and the nicotinic AChR subunits.

(i) The glcyine 48 kd subunit contains four putative transmembrane domains (M1–M4).
(ii) It contains two cysteine residues separated by 13 amino acids (analogous to *Torpedo* AChR α subunit Cys-128 and -142).
(iii) M2 is highly enriched in serine and threonine residues (containing eight such residues out of the 18 amino acids which comprise this domain).
(iv) A proline residue is present in the middle of M1.
(v) There are two proline residues located between M2 and M3.
(vi) There is one potential asparagine-linked glycosylation site located 38 residues from the amino terminus (similar to all of the neuronal AChR α and non-α subunits) (see *Figure 3*).
(vii) In terms of overall sequence identity, the glycine 48 kd is 21% identical to *Drosophila* AChR non-α subunit.

11.1.3 GABA receptors

It had been thought that the $GABA_A$ receptor was composed solely of an α (53 kd) and a β (57 kd) subunit in the stoichiometry $\alpha_2\beta_2$. It was further believed that benzodiazepines bound to the α subunit while agonists bound to the β subunit. Now that clones for these and other GABA receptor subunits have been isolated and expressed, neither the composition of the channel nor the function of the individual subunits is clear. Initially, a single GABA α cDNA clone and a single β cDNA clone were isolated from bovine brain and calf cerebral cortex cDNA libraries (35). Since then, at least four different subunit cDNAs have been isolated, termed α, β, γ and δ (309), and multiple subunit types have been isolated, termed α_1, α_2, α_3, etc. (310). Expression of any one of the subunits results in ligand-sensitive channels which are regulated by barbiturate and picrotoxin (309,311,312). Channels formed with α and β subunits possess many of the expected channel properties of $GABA_A$ receptors; however, they are partially impaired since they lack the expected sensitivity to benzodiazepines, and have a Hill coefficient for GABA of 1 rather than 2 (310). Channels formed with α, β and γ subunits have properties which more closely approximate those of previously characterized $GABA_A$ receptors; however, they still appear not to be fully functional (309).

Sequence analysis of the original α and β clones (35) revealed that these two subunits shared extensive similarities with AChR subunits and with the glycine receptor 48 kd subunit.

(i) GABA receptor α and β subunits each contain four putative transmembrane domains (M1 – M4).
(ii) They contain two cysteine residues separated by 13 amino acids.
(iii) M2 is highly enriched in serine and threonine residues (7 – 8 for each subunit).
(iv) A proline residue is present in the middle of M1.
(v) There are one or two proline residues located between M2 and M3.
(vi) The α subunit has two potential asparagine-linked glycosylation sites (one at position 11, similar to AChR neuronal subunits), and the β subunit has three (one at position 8, similar to AChR neuronal subunits and one at position 149, similar to muscle) (see *Figure 3*).
(vii) The GABA receptor α subunit is 18.5% identical to bovine AChR α subunit and the GABA receptor β subunit is 15% identical to bovine AChR α subunit.

As discussed in Section 5.3, evidence has accumulated suggesting that M1 and/or M2 might form the lining of the ion channel of the AChR. It is thus intriguing that conserved features among the AChRs regarding these two domains are also highly conserved in the glycine and GABA receptors. Other highly conserved structural features of all of these ligand-gated channels are the cysteine residues at positions 128 and 142. It has already been demonstrated for the *Torpedo* AChR α subunit that these two cysteine residues are disulfide-bonded to each other (172). Given the striking similarities just described, it is very likely that these Cys residues are similarly bonded in the other AChR subunits and in the GABA and glycine receptor subunits. The mere fact that these regions are so well conserved strongly implies that they perform essential structural and/or functional roles in ligand-gated receptor-channels. It is these structural, functional and primary sequence similarities which suggest that the nicotinic AChRs, GABA and glycine receptors all belong to a superfamily of neurotransmitter-gated ion channels.

11.2 Cloned voltage-gated channels

Members of three classes of voltage-gated channels have been cloned and sequenced. Na^+ channels have been isolated from *E.electricus* (42), rat brain (43), *Drosophila* (44,45) and rat skeletal muscle (46). A dihydropyridine receptor (Ca^{2+} channel) has been isolated from rabbit skeletal muscle (47,48), and potassium channels have been isolated from

Drosophila (49–55), mouse brain (56) and rat brain (57). As discussed below, sequence analysis has revealed many similarities among all the members of the voltage-gated channels.

11.2.1 Na^+ channels

Sodium channels purified from the electric organ of the eel *Electrophorus* (313,314) and from chick cardiac muscle (315) appear to be composed of a single large polypeptide of molecular mass 250 kd. In contrast, Na^+ channels purified from rat brain (316) and skeletal muscle (317) appear also to contain two or three smaller polypeptides with molecular masses ranging from 37 to 45 kd. From the amino acid sequence deduced from the cloned cDNA of the eel Na^+ channel (42), it was determined that the polypeptide is composed of 1820 amino acids with a calculated molecular mass of 208 321 daltons, having 10 potential asparagine-linked glycosylation sites. There are four highly homologous repeated units, each of which contains four (42), six (43), or possibly eight (318) putative transmembrane domains. Using the nomenclature of Noda *et al.* (42), the repeat units I–IV are composed of segments S1–S6. S4 has a very interesting composition and has been proposed to be the voltage sensor of the channel. Of the 21 amino acids which comprise this segment, five to seven of them have positively charged side chains, either lysine or arginine, located at every third position. Another very striking domain is the region between S6 of unit II and S1 of unit III. Of the 200 residues that comprise this region, 51 are aspartic and glutamic acid residues (25%). In one stretch, 12 of 14 amino acids are glutamic acids. It has been postulated that the gating current could be accounted for by the movement of groups of negatively and/or positively charged residues (319). One group has suggested that this highly negatively charged region might be involved in the gating current or possibly the inactivation of the channel (42). A different region has been proposed to be the possible site of inactivation (44). When the sequences from three other Na^+ channels were analyzed, a highly positively charged region was identified that contained six conserved lysine residues located between S6 of unit III and S1 of unit IV. This region has been postulated as the site of inactivation with the lysine residues as the sites of cleavage by trypsin whose action results in the elimination of inactivation (see refs in 44).

Analysis of the amino acid sequences deduced from the cloned DNAs encoding three distinct Na^+ channels from rat brain (43,320,321) and one from *Drosophila* (44) revealed that the overall scheme of four homologous repeated units each containing six putative transmembrane domains was conserved among the different Na^+ channels. None of these channels have signal or leader sequences and thus the amino termini are presumed to be located cytoplasmically. The calculated molecular masses for rat brain Na^+ channels I, II and III are 228 758 daltons (2009 residues), 227 840 daltons (2005 residues) and 221 375 daltons (1951 residues), respectively. There are six conserved potential asparagine-linked glycosylation sites among the three rat brain and one eel channels. The rat brain channels are similar to that of the eel, in that the S4 segments contain four to eight arginine and lysine residues located at every third position, and the regions between S6 of unit II and S1 of unit III (a domain of 215 amino acids) are very negatively charged containing 20% glutamic and aspartic acid residues. The degree of similarity among the rat brain channels is 85–87% and the similarity among the eel and rat brain channels is 61–62%.

Functional channels have not been expressed in oocytes after injection of the cRNA made from the eel clone; however, channels have been expressed from the individual rat brain clones (321–324). Although functional channels have been expressed, all of the channel properties appear not to have been reconstituted: the magnitude of the current produced from the type I channel is small, and although the currents produced by the type II and type III channels are comparable to those produced in oocytes after injection of unfractionated poly[A]$^+$ mRNA from rat brain, the type II channels have slower inactivation rates (321,323). That a smaller subunit(s) may be required in order to produce fully functional channels has been suggested by results reported by Auld et al. (321). This group was able to express Na$^+$ channels with inactivation kinetics identical to those produced from unfractionated poly[A]$^+$ mRNA but only after a low molecular weight fraction of rat brain mRNA was co-injected with the type II cRNA. These results are very intriguing and indicate that it may not be possible to thoroughly characterize the different sodium channels until clones for the $\beta1$ and/or $\beta2$ subunits are isolated.

11.2.2 Ca^{2+} channels

As discussed in Section 1.2.2, ion flux can be modulated by DHP in the L-type Ca^{2+} channel. Although the exact subunit composition of this channel is uncertain, DHP has been shown to bind to the 175 kd $\alpha1$ polypeptide. The cDNA for this polypeptide has been cloned from rabbit skeletal muscle and sequenced (47,48). The deduced polypeptide is composed of 1873 amino acids and has a calculated molecular mass of 212 018 daltons. There are four repeated homologous domains each composed of six putative transmembrane segments. The degree of similarity between rat brain Na$^+$ channel type II and the 175 kd $\alpha1$ polypeptide of the DHP receptor for the four repeated units is 62, 59, 60 and 61%, respectively. The overall similarity between these two channels is 55% with 29% of the residues being identical. As with the Na$^+$ channels, the DHP receptor $\alpha1$ polypeptide has no signal sequence. Other common features include:
(i) repeats of the five hydrophobic domains (S1, S2, S3, S5, S6);
(ii) the positively charged S4 domains contain five or six arginine and lysine residues located at every third position;

(iii) the 137 amino acid region between S6 of unit II and S1 of unit III contains 20% glutamic and aspartic acid residues.

Apparently, no functional channels have been expressed in oocytes after injection of cRNAs from the α1 subunit clones. However, in a very exciting report by Tanabe et al. (325), Ca^{2+} channel function could be rescued in cultured skeletal muscle cells from mice with muscular dysgenesis (mdg) after injection of a plasmid containing the α1 subunit cDNA. In addition, excitation – contraction coupling was also restored in these cells. This report is the first functional study suggesting that the cloned α1 polypeptide is indeed a Ca^{2+} channel. As mentioned in Section 1.2.2, sequence analysis of the α2 polypeptide revealed no homologies to other known proteins, and the function of this polypeptide is still unclear.

Although the channel-forming domains of Na^+ and Ca^{2+} channels appear to be contained within a single large polypeptide chain, there are several analogies that can be made with the multisubunit ligand-gated receptor-channels. The muscle-like AChRs are composed of five subunits each probably containing four transmembrane domains for a total of 20 membrane-spanning segments. The two types of voltage-gated channels have four repeated domains, each containing possibly six membrane-spanning segments for a total of 24 membrane crosses. Although the voltage-gated channels are not multisubunit complexes, the arrangement of the repeating units suggests that each unit is analogous to a subunit. Thus, the general structural features of ligand-gated and voltage-gated channels appear to be fairly similar in that they are large (250 kd) complexes, composed of multiple subunits or subunit-like domains, containing a total of 20 – 30 membrane crosses.

11.2.3 K^+ channels

The potassium-selective channels are a heterogeneous group of ion channels (40,64). The lack of a tissue source rich in a single population of channels and the lack of highly specific reagents has hindered the isolation of these channels. Several groups have therefore used a genetic approach to isolate them (49 – 55; see also Chapter 2). Using these clones, or information derived from them, K^+ channels have also been isolated from mouse (56) and rat (57) brains. Physiological analysis of *Shaker* mutants suggested that the *Shaker* locus of *D.melanogaster* encoded the voltage-sensitive K^+ channel called the A channel. The localization of the gene to specific polytene bands on the X chromosome and the availability of a cDNA clone which hybridized to this region allowed investigators to begin a chromosome 'walk' (65) in order to isolate the gene (49 – 51).

The deduced amino acid sequence of this K^+ channel (called ShA) is 616 amino acids encoding a polypeptide of molecular mass 70 200 daltons (50). Hydropathy analysis suggests six transmembrane domains (H1 – H6). However, one other putative transmembrane domain exists which is highly homologous to the positively charged S4 segment of the Na^+ and Ca^{2+} channels. The ShA S4 segment has seven arginine residues spaced every third amino acid and the region is 50% identical to S4 of the eel Na^+ channel, suggesting that S4 of ShA is the voltage-gating domain of this protein, as is thought to be the case for the Na^+ and Ca^{2+} channels. Like the other two voltage-gated channels, ShA does not contain a signal sequence. The overall structural similarities among the three channel types would suggest that the K^+ channel is composed of just one of the four repeating units observed in Na^+ and Ca^{2+} channels. Whether the naturally occurring channel is composed of multiple copies of this one polypeptide or composed of heterologous subunits is not yet known but functional channels have been expressed from single clones (see below). The observed similarities do suggest, however, that the channels all evolved from a single voltage-sensitive ancestor.

An intriguing observation was made by Schwarz et al. (53) and Kamb et al. (54) who determined that multiple K^+ channel mRNAs are encoded by alternative splicing at the *Shaker* locus. Possibly four to ten classes of K^+ channel cDNAs have been isolated. Thorough electrophysiological and pharmacological characterization of the expressed channels may reveal that they each encode a distinct K^+ channel or one subunit of a distinct channel. Microinjection into *Xenopus* oocytes of mRNA prepared *in vitro* using the ShA1 and ShB1 cDNAs as templates has demonstrated that functional channels can be formed with either of the two classes of transcript and, furthermore, that two types of A channels with different kinetic properties were formed (55). At this point, several possibilities for the expression of heterogeneous K^+ channels are feasible. Divergence among the K^+ channels could be achieved by expressing many different polypeptides (in part by using alternative splicing mechanisms) and forming channels either with homologous subunits, or with different combinations of heterologous A channel polypeptides, or with the A channel polypeptides in combination with other as yet uncharacterized subunits. The novel observation that multiple channel components can be expressed by alternative splicing of transcripts from a single gene suggests one very intriguing mechanism for generating channel diversity.

12. Conclusions

With the application of the techniques of molecular genetics to the study of receptors and channels, rapid progress has been made especially with the isolation of channels which have previously been poorly characterized (due to lack of sufficient quantities of material or high-specificity reagents). One of the most exciting discoveries to come from cloning and sequencing different channel types is that the ligand-gated receptor-channels appear to belong to one superfamily of channels and the voltage-gated channels appear to belong to another superfamily of channels. Although many structural and functional predictions are made with the isolation of each new channel, the conservation of homologous domains in channel isolates has made a significant impact on identifying these important regions.

We are still in the early stages of accumulating data on various channels, but the use of several different approaches in their isolation has greatly aided in this endeavor. The 'standard' approach for isolating several channel cDNAs has been first to obtain protein sequence information, prepare synthetic oligonucleotide probes, and then to screen cDNA libraries. Another successful variation on this approach is to screen expression libraries with specific antibodies (326). Other approaches which have been developed to isolate genes encoding proteins expressd on cell surfaces could also be used for the cloning of certain types of ion channel genes. DNA-mediated gene transfer techniques have been used in conjunction with a transient expression system in COS cells in which cells expressing the desired protein are isolated after adhesion to antibody-coated dishes. This approach has been used to isolate genes encoding nerve growth factor receptors (327,329) and the CD2 antigen (329), respectively. Two very different techniques have been used in order to isolate channels for which neither protein sequence information nor high-specificity reagents were available. Thus, *Drosophila* K^+ channels have been isolated using a genetic approach (described in Section 11.2.3 and Chapter 2) (49–55). In the other protocol, a cloning strategy was developed that relied on an electrophysiological assay (330). The serotonin 5-HT_{1C} receptor, when bound by ligand, couples to a G-protein and stimulates a Ca^{2+}-activated Cl^- channel. Lubbert *et al.* (330) constructed libraries in which the cDNAs were directionally inserted into single-stranded vectors in the antisense orientation. Single-stranded DNA was next prepared and hybridized with RNA for either depletion or selection procedures. Unhybridized (hybrid-depleted) or eluted (hybrid-selected) RNA was then injected into *Xenopus* oocytes and channel activity was assayed using electrophysiological measurements.

The channel which started all of this cloning activity was the nicotinic AChR from *Torpedo*. The cloning of its subunit cDNAs quickly led to the isolation of muscle AChRs from several sources and then to the isolation of neuronal AChRs. Given the complexity of the brain, the low level of expression of AChRs in brain tissue and the diversity of the neuronal AChRs, it was a major breakthrough to isolate neuronal AChRs using molecular genetic techniques and thereby characterize these channels. Whereas some people feared that there was little left to do once the first AChR had been cloned, in fact the field has expanded in several new directions, including the characterization of neuronal AChRs (Section 10), attempts to determine structure–function relationships using various mutagenesis techniques (Section 8), attempts to unravel the genomic organization of AChRs (Section 9), the isolation of novel subunits whose functions appear to have developmental implications (Section 6), and expression of AChRs in foreign environments to better address cell biological processes (Section 7).

There are still many unanswered questions concerning all aspects of receptor biology: how is gene expression regulated? How and where do subunits assemble? What is the mechanism of coupling between ligand binding and channel opening? What interactions exist between AChRs and cytoskeletal components or nerves? Answering these questions will be difficult but should be possible by combining standard pharmacological, electrophysiological and biochemical techniques with molecular genetics, *in-vitro* mutagenesis, and the application of gene expression systems. All of the varied and specific reagents which have contributed to the AChR becoming the best characterized receptor-channel (see Section 2) will continue to be essential for its analysis after the insults of molecular genetic techniques have been inflicted. Because this channel has been so thoroughly researched and because of the variety of excellent reagents available for its analysis, the nicotinic AChR may continue to be the exemplar for receptor-channels as well as the exemplar for protein complexes composed of heterologous subunits.

13. Acknowledgements

I thank Arthur Karlin (Columbia University), Fred Sigworth (Yale University School of Medicine), Marc Ballivet (University of Geneva), John Dani (Baylor College of Medicine) and William Green (Yale University School of Medicine) for their critical reading of the manuscript and helpful comments and suggestions. I also thank the following people for providing data and manuscripts before publication: Marc Ballivet, Jim Boulter (The Salk Institute), Steve Burden (Massachusetts Institute of Technology), Steve Heinemann (The Salk Institute), John Merlie (Washington University School of Medicine) and Mark Tanouye (California Institute of Technology). Work from my laboratory is supported by National Institutes of Health grants NS-21714 and HL-38156.

14. References

1. Karlin,A. (1980) Molecular properties of nicotinic acetylcholine receptors. In *The Cell Surface and Neuronal Function, Volume 6.* G.Poste, C.W.Cotman and G.L.Nicolson (eds), Elsevier/North-Holland Biomedical Press, Amsterdam, pp. 191–260.
2. Changeux,J.-P. (1981) The acetylcholine receptor: an allosteric membrane protein. In *Harvey Lectures, Volume 75.* Academic Press, New York, pp. 85–254.

3. Conti-Tronconi,B.M. and Raftery,M.A. (1982) The nicotinic cholinergic receptor: correlation of molecular structure with functional properties. *Annu. Rev. Biochem.*, **51**, 491–530.
4. Barrantes,F.J. (1983) Recent developments in the structure and function of the acetylcholine receptor. *Int. Rev. Neurobiol.*, **24**, 259–341.
5. Popot,J.-L. and Changeux,P.-J. (1984) Nicotinic receptor of acetylcholine: structure of an oligomeric integral membrane protein. *Physiol. Rev.*, **64**, 1162–1239.
6. Dolly,J.O. and Barnard,E.A. (1984) Nicotinic acetylcholine receptors: an overview. *Biochem. Pharm.*, **33**, 841–858.
7. Hucho,F. (1986) The nicotinic acetylcholine receptor and its ion channel. *Eur. J. Biochem.*, **158**, 211–226.
8. Maelicke,A. (1988) Structure and function of the nicotinic acetylcholine receptor. In *Handbook of Experimental Pharmacology: The Cholinergic Synapse, Volume 86.* V.P.Whittaker (ed.), Springer-Verlag, Berlin, pp. 267–300.
9. Hess,G.P., Cash,D.J. and Aoshima,H. (1983) Acetylcholine receptor-controlled ion translocation: chemical kinetic investigations of the mechanism. *Annu. Rev. Biophys. Bioeng.*, **12**, 443–473.
10. Udgaonkar,J.B. and Hess,G.P. (1986) Acetylcholine receptor kinetics: chemical kinetics. *J. Memb. Biol.*, **93**, 93–109.
11. Hess,G.P., Udgaonkar,J.B. and Olbricht,W.L. (1987) Chemical kinetic measurements of transmembrane processes using rapid reaction techniques: acetylcholine receptor. *Annu. Rev. Biophys. Biophys. Chem.*, **16**, 507–534.
12. Peper,K., Bradley,R.J. and Dreyer,F. (1982) The acetylcholine receptor at the neuromuscular junction. *Physiol. Rev.*, **62**, 1271–1340.
13. Sakmann,B. and Neher,E. (1984) Patch-clamp techniques for studying ionic channels in excitable membranes. *Annu. Rev. Physiol.*, **46**, 455–472.
14. Colquhoun,D. and Sakmann,B. (1985) Fast events in single-channel currents activated by acetylcholine and its analogues at the frog muscle end-plate. *J. Physiol.*, **369**, 501–557.
15. Adams,P.R. (1987) Transmitter action at endplate membrane. In *Neurology and Neurobiology, Volume 23.* M.M.Salpeter (ed.), Alan R.Liss, Inc., New York, pp. 317–359.
16. Fambrough,D.M. (1979) Control of acetylcholine receptors in skeletal muscle. *Physiol. Rev.*, **59**, 165–227.
17. Fambrough,D.M. (1983) Biosynthesis and intracellular transport of acetylcholine receptors. *Methods Enzymol.*, **96**, 331–352.
18. Merlie,J.P. and Smith,M.M. (1986) Synthesis and assembly of acetylcholine receptor, a multisubunit membrane glycoprotein. *J. Membr. Biol.*, **91**, 1–10.
19. Dennis,M.J. (1981) Development of the neuromuscular junction: inductive interactions between cells. *Annu. Rev. Neurosci.*, **4**, 43–68.
20. Schuetze,S.M. and Role,L.W. (1987) Developmental regulation of nicotinic acetylcholine receptors. *Annu. Rev. Neurosci.*, **10**, 403–457.
21. Salpeter,M.M. (1987) Development and neural control of the neuromuscular junction and of the junctional acetylcholine receptor. In *Neurology and Neurobiology, Volume 23.* M.M.Salpeter (ed.), Alan R.Liss, Inc., New York, pp. 55–115.
22. Bloch,R.J. and Pumplin,D.W. (1988) Molecular events in synaptogenesis: nerve–muscle adhesion and postsynaptic differentiation. *Am. J. Physiol.*, **254**, C345–C364.
23. Brehm,P. and Henderson,L. (1988) Regulation of acetylcholine receptor function during development of skeletal muscle. *Dev. Biol.*, **129**, 1–11.
24. Vincent,A. (1980) Immunology of acetylcholine receptors in relation to myasthenia gravis. *Physiol. Rev.*, **60**, 754–824.
25. Lindstrom,J. (1985) Immunobiology of myasthenia gravis, experimental autoimmune myasthenia gravis, and lambert–eaton syndrome. *Annu. Rev. Immunol.*, **3**, 109–131.
26. Engel,A.G. (1987) Molecular biology of endplate diseases. In *The Vertebrate Neuromuscular Junction.* M.M.Salpeter (ed.), Alan R.Liss, Inc., New York, pp. 361–424.
27. Anholt,R., Lindstrom,J. and Montal,M. (1984) The molecular basis of neurotransmission: structure and function of the nicotinic acetylcholine receptor. In *The Enzymes of Biological Membranes, Volume 3.* A.Martonosi (ed.), Plenum Press, New York, pp. 335–401.

28. Stroud,R.M. and Finer-Moore,J. (1985) Acetylcholine receptor structure, function, and evolution. *Annu. Rev. Cell Biol.*, **1**, 369–401.
29. Maelicke,A. and Prinz,H. (1983) The acetylcholine receptor-ion channel complex: linkage between binding and response. In *Modern Cell Biology, Volume 1*. E.H.Satir (ed.), Alan R.Liss, New York, pp. 171–197.
30. Claudio,T. (1986) Recombinant DNA technology in the study of ion channels. *Trends Pharmacol. Sci.*, **7**, 308–312.
31. Anderson,D.J. (1987) Molecular biology of the acetylcholine receptor: structure and regulation of biogenesis. In *Neurology and Neurobiology, Volume 23*. M.M.Salpeter (ed.), Alan R.Liss, Inc., New York, pp. 285–315.
32. Sumikawa,K., Houghton,M., Emtage,J.S., Richards,B.M. and Barnard,E.A. (1981) Active multi-subunit ACh receptor assembled by translation of heterologous mRNA in *Xenopus* oocytes. *Nature*, **292**, 862–864.
33. Doolittle,R.F. (1981) Similar amino acid sequences: chance or common ancestry? *Science*, **214**, 149–159.
34. Grenningloh,G., Rienitz,A., Schmitt,B., Methfessel,C., Zensen,M., Beyreuther,K., Gundelfinger,E.D. and Betz,H. (1987) The strychnine-binding subunit of the glycine receptor shows homology with nicotinic acetylcholine receptors. *Nature*, **328**, 215–220.
35. Schofield,P.R., Darlison,M.G., Fujita,N., Burt,D.R., Stephenson,F.A., Rodriguez,H., Rhee,L.M., Ramachandran,J., Reale,V., Glencorse,T.A., Seeburg,P.H. and Barnard,E.A. (1987) Sequence and functional expression of the $GABA_A$ receptor shows a ligand-gated receptor super-family. *Nature*, **328**, 221–227.
36. Jones,S.W. (1987) Presynaptic mechanisms at vertebrate neuromuscular junctions. In *Neurology and Neurobiology, Volume 23*. M.M.Salpeter (ed.), Alan R.Liss, Inc., New York, pp. 187–245.
37. Salpeter,M.M. (1987) Vertebrate neuromuscular junctions: general morphology, molecular organization, and functional consequences. In *Neurology and Neurobiology, Volume 23*. M.M.Salpeter (ed.), Alan R.Liss, Inc., New York, pp. 1–54.
38. Betz,H. (1987) Biology and structure of the mammalian glycine receptor. *Trends Neurosci.*, **10**, 113–117.
39. Olsen,R.W. and Venter,J.C. (eds) (1986) *Benzodiazepine/GABA Receptors and Chloride Channels: Structural and Functional Properties.* Receptor Biochemistry and Methodology, Volume 5. Alan R.Liss, Inc., New York,
40. Hille,B. (1984) *Ionic Channels of Excitable Membranes.* Sinauer Assoc. Inc., Sunderland, MA.
41. Catterall,W.A. (1988) Structure and function of voltage-sensitive ion channels. *Science*, **242**, 50–61.
42. Noda,M., Shimizu,S., Tanabe,T., Takai,T., Kayano,T., Ikeda,T., Takahashi,H., Nakayama,H., Kanaoka,Y., Minamino,N., Kangawa,K., Matsuo,H., Raftery,M.A., Hirose,T., Inayama,S., Hayashida,H., Miyata,T. and Numa,S. (1984) Primary structure of *Electrophorus electricus* sodium channel deduced from cDNA sequence. *Nature*, **312**, 121–127.
43. Noda,M., Ikeda,T., Kayano,T., Suzuki,H., Takeshima,H., Kurasaki,M., Takahashi, H. and Numa,S. (1986) Existence of distinct sodium channel messenger RNAs in rat brain. *Nature*, **320**, 188–192.
44. Salkoff,L., Butler,A., Wei,A., Scavarda,N., Giffen,K., Ifune,C., Goodman,R. and Mandel,G. (1987) Genomic organization and deduced amino acid sequence of a putative sodium channel gene in *Drosophila*. *Science*, **237**, 744–749.
45. Ganetzky,B. and Loughney,K. (1988) The *Drosophila para* locus is a sodium channel structural gene. *Soc. Neurosci. Abs.*, **14**, 598 (241.7).
46. Trimmer,J.S., Agnew,W.S., Tomiko,S.A., Crean,S.M., Sheng,Z., Kallen,R., Barchi,R.L., Cooperman,S.S., Goodman,R.H. and Mandel,G. (1988) Isolation of cDNA clones encoding a full length rat skeletal muscle sodium channel. *Soc. Neurosci. Abs.*, **14**, 598 (241.8).
47. Tanabe,T., Takeshima,H., Mikami,A., Flockerzi,V., Takahashi,H., Kangawa,K., Kojima,M., Matsuo,H., Hirose,T. and Numa,S. (1987) Primary structure of the receptor for calcium blockers from skeletal muscle. *Nature*, **328**, 313–318.
48. Ellis,S.B., Williams,M.E., Ways,N.R., Brenner,R., Sharp,A.H., Leung,A.T., Campbell,K.P., McKenna,E., Koch,W.J., Hui,A., Schwartz,A. and Harpold,M.M. (1988) Sequence and expression of mRNAs encoding the α_1 and α_2 subunits of a

DHP-sensitive calcium channel. *Science,* **241**, 1661–1664.
49. Papazian,D.M., Schwarz,T.L., Tempel,B.L., Jan,Y.N. and Jan,L.Y. (1987) Cloning of genomic and complementary DNA from *Shaker*, a putative potassium channel gene from *Drosophila*. *Science,* **237**, 749–753.
50. Temple,B.L., Papazian,D.M., Schwarz,T.L., Jan,Y.N. and Jan,L.Y. (1987) Sequence of a probable potassium channel component encoded at *Shaker* locus of *Drosophila*. *Science,* **237**, 770–775.
51. Kamb,A., Iverson,L.E. and Tanouye,M.A. (1987) Molecular characterization of *Shaker*, a *Drosophila* gene that encodes a potassium channel. *Cell,* **50**, 405–413.
52. Baumann,A., Krah-Jentgens,I., Muller,R., Muller-Holtkamp,F., Seidel,R., Kecskemethy,N., Casal,J., Ferrus,A. and Pongs,O. (1987) Molecular organization of the maternal effect region of the *Shaker* complex of *Drosophila*: characterization of an I_A channel transcript with homology to vertebrate Na$^+$ channel. *EMBO J.,* **6**, 3419–3429.
53. Schwarz,T.L., Tempel,B.L., Papazian,D.M., Jan,Y.N. and Jan,L.Y. (1988) Multiple potassium-channel components are produced by alternative splicing at the *Shaker* locus in *Drosophila*. *Nature,* **331**, 137–142.
54. Kamb,A., Tseng-Crank,J. and Tanouye,M.A. (1988) Multiple products of the *Drosophila Shaker* gene contribute to potassium channel diversity. *Neuron,* **1**, 421–430.
55. Timpe,L.C., Schwarz,T.L., Tempel,B.L., Papazian,D.M., Jan,Y.N. and Jan,L.Y. (1988) Expression of functional potassium channels from *Shaker* cDNA in *Xenopus* oocytes. *Nature,* **331**, 143–145.
56. Tempel,B.L., Urban,J., Dorsa,D.M., Timpe,L.C., Jan,Y.N. and Jan,L.Y. (1988) Cloning and expression of a probable potassium channel gene from mouse brain. *Soc. Neurosci. Abs,* **14**, 455 (186.6).
57. Kaczmarek,L.K., Boyle,M.B., Blumenthal,E. and Marshall,J. (1988) Cloning of a shaker potassium channel homologue from rat brain. *Soc. Neurosci. Abs.,* **14**, 455 (186.7).
58. Catterall,W.A. (1986) Molecular properties of voltage-sensitive sodium channels. *Annu. Rev. Biochem.,* **55**, 953–985.
59. Barchi,R.L. (1988) Probing the molecular structure of the voltage-dependent sodium channel. *Annu. Rev. Neurosci.,* **11**, 455–495.
60. Trimmer,J.S. and Agnew,W.S. (1989) Molecular diversity of voltage-sensitive Na channels. *Annu. Rev. Physiol.,* **51**, 401–418.
61. Reuter,H., Porzig,H., Kokubun,S. and Prud'hom,B. (1985) 1,4-dihydropyridines as tools in the study of Ca^{2+} channels. *Trends Neurosci.,* **8**, 396–400.
62. Articles in: (1988) Special issue—Calcium and neuronal excitability. *Trends Neurosci.,* **11**, 415–469.
63. Catterall,W.A., Seagar,M.J. and Takahashi,M. (1988) Molecular properties of dihydropyridine-sensitive calcium channels in skeletal muscle. *J. Biol. Chem.,* **263**, 3535–3538.
64. Latorre,R. and Miller,C. (1983) Conduction and selectivity in potassium channels. *J. Membr. Biol.,* **71**, 11–30.
65. Salkoff,L.B. and Tanouye,M.A. (1986) Genetics of ion channels. *Physiol. Rev.,* **66**, 301–329.
66. Chagas,C. and Paes de Carvalho,A. (eds) (1961) *Bioelectrogenesis*. Elsevier Publishing Co., Amsterdam.
67. Lee,C.Y. (1972) Chemistry and pharmacology of polypeptide toxins in snake venoms. *Annu. Rev. Pharmacol.,* **12**, 265–286.
68. Cartaud,J. (1980) A critical re-evaluation of the structural organization of the excitable membrane in *Torpedo marmorata* electric organ. In *Ontogenesis and Functional Mechanism of Peripheral Synapses*. J.Taxi (ed.), Elsevier, Amsterdam, pp. 199–210.
69. Ross,M.J., Klymkowsky,M.W., Agard,D.A. and Stroud,R.M. (1977) Structural studies of a membrane-bound acetylcholine receptor from *T.californica. J. Mol. Biol.,* **116**, 635–659.
70. Zingsheim,H.P., Neugebauer,D.-C., Frank,J., Hanicke,W. and Barrantes,F.J. (1982) Dimeric arrangement and structure of the membrane-bound acetylcholine receptor studied by electron microscopy. *EMBO J.,* **1**, 541–547.
71. Wise,D., Karlin,A. and Schoenborn,B.P. (1979) Analysis by low-angle neutron scattering of the structure of the acetylcholine receptor from *Torpedo californica*

in detergent solution. *Biophys. J.*, **28**, 473–496.
72. Heuser,J.E. and Salpeter,S.R. (1979) Organization of acetylcholine receptors in quick-frozen, deep-etched, and rotary-replicated *Torpedo* postsynaptic membrane. *J. Cell Biol.*, **82**, 150–173.
73. Brisson,A. and Unwin,P.N.T. (1985) Quaternary structure of the acetylcholine receptor. *Nature*, **315**, 474–477.
74. Maeno,T., Edwards,C. and Ankaru,M.(1977) Permeability of the end-plate membrane activated by acetylcholine to some organic cations. *J. Neurobiol.*, **8**, 173–184.
75. Blanchard,S.G. and Raftery,M.A. (1979) Identification of the polypeptide chains in *Torpedo californica* electroplax membranes that interact with a local anesthetic analog. *Proc. Natl. Acad. Sci. USA*, **76**, 81–85.
76. Oswald,R.E. and Changeux,J.-P. (1981) Selective labeling of the δ subunit of the acetylcholine receptor by a covalent local anesthetic. *Biochemistry*, **20**, 7166–7174.
77. Hucho,F., Oberthur,W. and Lottspeich,F. (1986) The ion channel of the nicotinic acetylcholine receptor is formed by the homologous helices M II of the receptor subunits. *FEBS Letts.*, **205**, 137–142.
78. Giraudat,J., Dennis,M., Heidemann,T., Chang,J.-Y. and Changeux,J.-P. (1986) Structure of the high-affinity binding site for noncompetitive blockers of the acetylcholine receptor: serine-262 of the δ subunit is labeled by [^{3}H]chlorpromazine. *Proc. Natl. Acad. Sci. USA*, **83**, 2719–2723.
79. Kaldany,R.-R. and Karlin,A. (1983) Reaction of quinacrine mustard with the acetylcholine receptor from *Torpedo californica*. Functional consequences and sites of labeling. *J. Biol. Chem.*, **258**, 6232–6242.
80. Muhn,P. and Hucho,F. (1983) Covalent labeling of the acetylcholine receptor from *Torpedo* electric tissue with the channel blocker [^{3}H]triphenylmethylphosphonium by ultraviolet irradiation. *Biochemistry*, **22**, 421–425.
81. Lauffer,L., Weber,K.-H. and Hucho,F. (1979) Acetylcholine receptor. Binding properties and ion permeability response after covalent attachment of the local anesthetic quinacrine. *Biochim. Biophys. Acta*, **587**, 42–48.
82. Karlin,A. (1983) Anatomy of a receptor. *Neurosci. Commentaries*, **1**, 111–123.
83. Taylor,P., Brown,R.D. and Johnson,D.A. (1983) The linkage between ligand occupation and response of the nicotinic acetylcholine receptor. In *Current Topics in Membrane Transport. Volume 18.* A.Kleinzeller and B.R.Martin (eds), Academic Press, New York, pp. 407–444.
84. Changeux,J.-P., Devillers-Thiery,A. and Chemouilli,P. (1984) Acetylcholine receptor: an allosteric protein. *Science*, **225**, 1335–1345.
85. Karlin,A., Kao,P.N. and DiPaola,M. (1986) Molecular pharmacology of the nicotinic acetylcholine receptor. *Trends Pharmacol. Sci.*, **7**, 304–308.
86. Adams,D.J., Dwyer,T.M. and Hille,B. (1980) The permeability of endplate channels to monovalent and divalent metal cations. *J. Gen. Physiol.*, **75**, 493–510.
87. Neher,E. and Sakmann,B.(1976) Single channel currents recorded from membrane of denervated frog muscle fibers. *Nature*, **260**, 799–802.
88. Hamill,O.P., Marty,A., Neher,E., Sakmann,B. and Sigworth,F.J. (1981) Improved patch-clamp techniques for high-resolution current reading from cells and cell-free membrane patches. *Pfluger's Arch.*, **391**, 85–100.
89. Corey,D.P. (1983) Patch clamp: current excitement in membrane physiology. *Neurosci. Commentaries*, **1**, 99–110.
90. Sigworth,F.J. (1986) The patch clamp is more useful than anyone had expected. *Fed. Proc.*, **45**, 2673–2677.
91. Usdin,T.B. and Fischbach,G.D. (1986) Purification and characterization of a polypeptide from chick brain that promotes the accumulation of acetylcholine receptors in chick myotubes. *J. Cell Biol.*, **103**, 493–507.
92. Rubin,L.L. and Barald,K.F. (1983) Neuromuscular development in tissue culture. In *Somatic and Autonomic Nerve–Muscle Interactions*. G.Burnstock, G.Vrbova and R.O'Brien (eds), Elsevier Science Pub., B.V., pp. 109–151.
93. Steinbach,J.H. and Block,R. (1986) Control of acetylcholine receptor distribution in vertebrate skeletal muscle. In *Receptors in Cell Recognition and Differentiation*. R.M.Gorczynski and G.B.Price (eds), Academic Press, New York, pp. 183–213.
94. Edwards,C. and Frisch,H.L. (1976) A model for the localization of acetylcholine receptors at the muscle endplate. *J. Neurobiol.*, **7**, 377–381.

95. Fraser,S.E. and Poo,M.-m. (1982) Development, maintenance and modulation of patterned membrane topology: models based on the acetylcholine receptor. *Curr. Top. Devel. Biol.*, **17**, 77–100.
96. Englander,L.L. and Rubin,L.L. (1987) Acetylcholine receptor clustering and nuclear movement in muscle fibers in culture. *J. Cell Biol.*, **104**, 87–95.
97. Merlie,J.P. and Sanes,J.R. (1985) Concentration of acetylcholine receptor mRNA in synaptic regions of adult muscle fibers. *Nature*, **317**, 66–68.
98. Nitkin,R.M., Smith,M.A., Magill,C., Fallon,J.R., Yao,Y.-M.M., Wallace,B.G. and McMahan,U.J. (1987) Identification of agrin, a synaptic organizing protein from *Torpedo* electric organ. *J. Cell Biol.*, **105**, 2471–2478.
99. Block,R.J. and Morrow,J.S. (1987) A unique β-spectrin associated with clustered acetylcholine receptors. *J. Cell Biol.*, **105**, 290a (1632).
100. Daniels,M.P., Olek,A.J., Zeng,F.J., Krikorian,J.G. and Block,R.J. (1987) Vinculin, β-actinin and 43K protein associated with newly formed acetylcholine receptor aggregates. *J. Cell Biol.*, **105**, 62a (344).
101. Yamaguchi,T.P. and Connolly,J.A. (1987) Role of cytoskeletal associated proteins in acetylcholine receptor cluster stabilization in chick myotubes. *J. Cell Biol.*, **105**, 63a (350).
102. Peng,H.B., Chen,Q.-M., Rochlin,W., Tobler,M., Turner,C.E. and Burridge,K. (1987) The distribution of talin at ACh receptor clusters in muscle cells. *J. Cell Biol.*, **105**, 63a (345).
103. Froehner,S.C. (1986) The role of the postsynaptic cytoskeleton in AChR organization. *Trends Neurosci.*, **9**, 37–41
104. Tsui,H.-C.T., Carr,C., Cohen,J.B. and Fischbach,G.D. (1987) Acetylcholine receptor clusters are not invariably associated with 43 K protein at early stages of chick muscle development. *Soc. Neurosci. Abs*, **13**, 797 (223.9).
105. Cartaud,J., Kordeli,E., Nghiem,H.O. and Changeux,J.-P. (1987) Assembly of acetylcholine receptor-rich membrane domains in *Torpedo marmorata* electrocyte during development. *J. Cell Biol.*, **105**, 145a (813).
106. Brockes,J.P. and Hall,Z.W. (1975) Acetylcholine receptors in normal and denervated rat diaphragm muscle. II. Comparison of junctional and extrajunctional receptors. *Biochemistry*, **14**, 2100–2105.
107. Hall,Z.W., Roison,M.-P., Gu,Y. and Gorin,P.D. (1983) A developmental change in the immunological properties of acetylcholine receptors at the rat neuromuscular junction. *Cold Spring Harbor Symp. Quant. Biol.*, **48**, 101–108.
108. Teichberg,V.I. and Changeux,J.-P. (1976) Presence of two forms of acetylcholine receptor with different isoelectric points in the electric organ of *Electrophorous electricus* and their catalytic interconversion *in vitro*. *FEBS Letts.*, **67**, 264–268.
109. Almon,R.R. and Appel,S.H. (1975) Interaction of myasthenia serum globulin with the acetylcholine receptor. *Biochim. Biophys. Acta*, **393**, 66–77.
110. Weinberg,C.B. and Hall,Z.W. (1979) Antibodies from patients with myasthenia gravis recognize determinants unique to extracellular acetylcholine receptors. *Proc. Natl. Acad. Sci. USA*, **76**, 504–518.
111. Kullberg,R.W., Brehm,P. and Steinbach,J.H. (1981) Nonjunctional acetylcholine receptor channel open time decreases during development of *Xenopus* muscle. *Nature*, **289**, 411–413.
112. Brehm,P., Kidokoro,Y. and Moody-Corbett,F. (1984) Acetylcholine receptor channel properties during development of *Xenopus* muscle cells in culture. *J. Physiol.*, **357**, 203–217.
113. Schuetze,S.M. and Vicini,S. (1984) Neonatal denervation inhibits the normal postnatal decrease in endplate channel open time. *J. Neurosci.*, **4**, 2297–2302.
114. Brehm,P. and Kullberg,R. (1987) Acetylcholine receptor channels on adult mouse skeletal muscle are functionally identical in synaptic and nonsynaptic membrane. *Proc. Natl. Acad. Sci. USA*, **84**, 2550–2554.
115. Patrick,P. and Lindstrom,J. (1973) Autoimmune response to acetylcholine receptor. *Science*, **180**, 871–872.
116. Claudio,T. and Raftery,M.A. (1980) Is experimental autoimmune myasthenia gravis induced only by acetylcholine receptors? *J. Immunol.*, **124**, 1130–1140.
117. Fenichel,G.M. (ed.) (1985) *Clinical Neurology and Neurosurgery Monographs: Neonatal Neurology*. Churchill Livingstone, Inc., New York.

118. Newsome-Davis,J. (1986) Diseases of the neuromuscular junction. In *Diseases of the Nervous System, Clinical Neurobiology, Volume I.* R.K.Asbury, M.Guy, W.McKhann and I.McDonald (eds), Ardmore Medical Books, Philadelphia, pp. 269–282.
119. Drachman,D.B. (ed.) (1987) Myasthenia gravis: biology and treatment. *Ann. NY Acad. Sci.,* **505**, 1–909.
120. Huganir,R.L. and Racker,E. (1980) Endogenous and exogenous proteolysis of the acetylcholine receptor from *Torpedo californica. J. Supramol. Struct.,* **14**, 13–19.
121. Mendez,B., Valenzuela,P., Martial,J.A. and Baxter,J.D. (1980) Cell-free synthesis of acetylcholine receptor polypeptides. *Science,* **209**, 695–697.
122. Anderson,D.J. and Blobel,G. (1981) *In vitro* synthesis, glycosylation and membrane insertion of the four subunits of *Torpedo* acetylcholine receptor. *Proc. Natl. Acad. Sci. USA,* **78**, 5598–5602.
123. Barnard,E.A., Miledi,R. and Sumikawa,K. (1982) Translation of exogenous messenger RNA coding for nicotinic acetylcholine receptors produces functional receptors in *Xenopus* oocytes. *Proc. R. Soc. Lond. B,* **215**, 241–246.
124. Raftery,M.A., Hunkapiller,M.W., Strader,C.D. and Hood,L.E. (1980) Acetylcholine receptor: complex of homologous subunits. *Science,* **280**, 1454–1457.
125. Claudio,T. and Raftery,M.A. (1977) Immunological comparison of acetylcholine receptors and their subunits from species of electric ray. *Arch. Biochem. Biophys.,* **181**, 484–489.
126. Sumikawa,K., Houghton,M., Smith,J.C., Richards,B.M. and Barnard,E.A. (1982) The molecular cloning and characterisation of cDNA coding for the α subunit of the acetylcholine receptor. *Nucleic Acids Res.,* **10**, 5809–5822.
127. Noda,M., Takahashi,H., Tanabe,T., Toyosato,M., Furutani,Y., Hirose,T., Asai,M., Inayama,S., Miyata,T. and Numa,S. (1982) Primary structure of α-subunit precursor of *Torpedo californica* acetylcholine receptor deduced from cDNA sequence. *Nature,* **299**, 793–797.
128. Noda,M., Takahashi,H., Tanabe,T., Toyosato,M., Kikykotani,S., Hirose,T., Asai,M., Takashima,H., Inayama,S., Miyata,T. and Numa,S. (1982) Primary structures of β- and δ-subunit precursors of *Torpedo californica* acetylcholine receptor deduced from cDNA sequence. *Nature,* **299**, 793–797.
129. Noda,M., Takahashi,H., Tanabe,T., Toyosato,M., Kikyotani,S., Furutani,Y., Hirose,T., Takashima,H., Inayama,S., Miyata,T. and Numa,S. (1983) Structural homology of *Torpedo californica* acetylcholine receptor subunits. *Nature,* **302**, 528–532.
130. Ballivet,M., Patrick,J., Lee,J. and Heinemann,S. (1982) Molecular cloning of cDNA coding for the γ subunit of *Torpedo* acetylcholine receptor. *Proc. Natl. Acad. Sci. USA,* **79**, 4466–4470.
131. Giraudat,J., Devillers-Thiery,A., Auffray,C., Rougeon,F. and Changeux,J.-P. (1982) Identification of a cDNA clone coding for the acetylcholine receptor binding subunit of *Torpedo marmorata* acetylcholine receptor. *EMBO J.,* **1**, 713–717.
132. Hershey,N.D., Noonan,D.J., Mixter,K.S., Claudio,T. and Davidson,T. (1983) Structure and expression of genomic clones coding for the δ-subunit of the *Torpedo* acetylcholine receptor. *Cold Spring Harbor Symp. Quant. Biol.,* **48**, 79–82.
133. Claudio,T., Ballivet,M., Patrick,J. and Heinemann,S. (1983) Nucleotide and deduced amino acid sequences of *Torpedo californica* acetylcholine receptor γ subunit. *Proc. Natl. Acad. Sci. USA,* **80**, 1111–1115.
134. Devillers-Thiery,A., Giraudat,J., Bentaboulet,M. and Changeux,J.-P. (1983) Complete mRNA coding sequence of the acetylcholine binding α-subunit of *Torpedo marmorata* acetylcholine receptor: a model for the transmembrane organization of the polypeptide chain. *Proc. Natl. Acad. Sci. USA,* **80**, 2067–2071.
135. Claudio,T., Palazzolo,M.J. and Axel,R. (1983) Introduction of *Torpedo* acetylcholine receptor genes into mammalian cells. *Soc. Neurosci. Abs.,* **9**, 620 (184.3).
136. Claudio,T. (1987) Stable expression of transfected *Torpedo* acetylcholine receptor α subunits in mouse fibroblast L cells. *Proc. Natl. Acad. Sci. USA,* **84**, 5967–5971.
137. Setzer,D.R., McGrogan,M., Numberg,J.H. and Schimke,R.T. (1980) Size heterogeneity in the 3' end of dihydrofolate reductase messenger RNAs in mouse cells. *Cell,* **22**, 361–370.
138. Proudfoot,N.J. and Brownlee,G.G. (1976) 3' Non-coding region sequences in eukaryotic messenger RNA. *Nature,* **263**, 211–214.

139. Conti-Tronconi,B.M., Hunkapiller,M.W. and Raftery,M.A. (1984) Molecular weight and structural nonequivalence of the mature α subunits of *Torpedo californica* acetylcholine receptor. *Proc. Natl. Acad. Sci. USA,* **81**, 2631–2634.
140. Anderson,D.J., Walter,P. and Blobel,G. (1982) Signal recognition protein is required for the integration of acetylcholine receptor δ subunit, a transmembrane glycoprotein, into the endoplasmic reticulum membrane. *J. Cell Biol.,* **93**, 501–506.
141. Lingappa,V.R., Katz,F.N., Lodish,H.F. and Blobel,G. (1978) A signal sequence for the insertion of a transmembrane glycoprotein. *J. Biol. Chem.,* **253**, 8667–8670.
142. Vandlen,R.L., Wu,W.C.-S., Eisenach,J.C. and Raftery,M.A. (1979) Studies of the composition of purified *Torpedo californica* acetylcholine receptor and of its subunits. *Biochemistry,* **10**, 1845–1854.
143. Lindstrom,J., Merlie,J. and Yogeeswaran,G. (1979) Biochemical properties of acetylcholine receptor subunits from *Torpedo californica*. *Biochemistry,* **18**, 4465–4470.
144. Nomoto,H., Takahashi,N., Nagaki,Y., Endo,S., Arata,Y. and Hayashi,K. (1986) Carbohydrate structures of acetylcholine receptor from *Torpedo californica* and distribution of oligosaccharides among the subunits. *Eur. J. Biochem.,* **157**, 133–242.
145. Takai,T., Noda,M., Furutani,Y., Takahashi,H., Notake,M., Shimizu,S., Kayano,T., Tanabe,T., Tanaka,K., Hirose,T., Inayama,S. and Numa,S. (1984) Primary structure of γ subunit precursor of calf-muscle acetylcholine receptor deduced from the cDNA sequence. *J. Biochem.,* **143**, 109–115.
146. Nef,P., Mauron,A., Stalder,R., Alliod,C. and Ballivet,M. (1984) Structure, linkage, and sequence of the two genes encoding the δ and γ subunits of the nicotinic acetylcholine receptor. *Proc. Natl. Acad. Sci. USA,* **81**, 7975–7979.
147. Shibahara,S., Kubo,T., Perski,P., Takahashi,H., Noda,M. and Numa,S. (1985) Cloning and sequence analysis of human genomic DNA encoding γ subunit precursor of muscle acetylcholine receptor. *Eur. J. Biochem.,* **146**, 15–22.
148. Boulter,J., Evans,K., Goldman,D., Martin,G., Treco,D., Heinemann,S. and Patrick,J. (1986) Isolation of a cDNA clone coding for a possible neural nicotinic acetylcholine receptor α-subunit. *Nature,* **319**, 368–374.
149. Yu,L., LaPolla,R.J. and Davidson,N. (1986) Mouse muscle nicotinic acetylcholine receptor γ subunit: cDNA sequence and gene expression. *Nucleic Acids Res.,* **14**, 3539–3555.
150. Baldwin,T.J., Yoshihara,C.M., Blackmer,K., Kintner,C.R. and Burden,S.J. (1988) Regulation of acetylcholine receptor transcript expression during development in *Xenopus laevis*. *J. Cell Biol.,* **106**, 469–478.
151. Claudio,T., Paulson,H.L., Green,W.N., Ross,A.F., Hartman,D.S. and Hayden,D. (1989) Fibroblasts transfected with *Torpedo* acetylcholine receptor β, γ and δ subunit cDNAs express functional AChRs when infected with a retroviral α-recombinant. *J. Cell Biol.,* **108**, 2277–2290.
152. Mishina,M., Tobimatsu,T., Tanaka,K., Fujita,Y., Fukuda,K., Kurasake,M., Takahashi,H., Morimoto,Y., Hirose,T., Inayama,S., Takahashi,T., Kuno,M. and Numa,S. (1985) Location of functional regions of acetylcholine receptor α-subunit by site-directed mutagenesis. *Nature,* **313**, 364–369.
153. Kao,P.N., Dwork,A.J., Kaldany,R.-R.J., Silver,M.L., Wideman,J., Stein,S. and Karlin,A. (1984) Identification of the α subunit half-cystine specifically labeled by an affinity reagent for the acetylcholine receptor binding site. *J. Biol. Chem.,* **259**, 11662–11665.
154. Huganir,R.L. and Greengard,P. (1983) cAMP-dependent protein kinase phosphorylates the nicotinic acetylcholine receptor. *Proc. Natl. Acad. Sci. USA,* **80**, 1130–1134.
155. Huganir,R.L. (1988) Regulation of the nicotinic acetylcholine receptor by protein phosphorylation. *Curr. Top. Membr. Transp.,* **33**, 147–163.
156. Safran,A., Sagi-Eisenberg,R., Neumann,D. and Fuchs,S. (1987) Phosphorylation of the acetylcholine receptor by protein kinase C and identification of the phosphorylation site within the receptor δ subunit. *J. Biol. Chem.,* **262**, 10506–10510.
157. Huganir,R.L., Miles,K. and Greengard,P. (1984) Phosphorylation of the nicotinic acetylcholine receptor by an endogenous tyrosine-specific kinase. *Proc. Natl. Acad. Sci. USA,* **81**, 6968–6972.
158. Huganir,R.L. and Greengard,P. (1987) Regulation of receptor function by protein phosphorylation. *Trends Pharm. Sci.,* **8**, 472–477.

159. Huganir,R.L., Delcour,A.H., Greengard,P. and Hess,G.P. (1986) Phosphorylation of the nicotinic acetylcholine receptor regulates its rate of desensitization. *Nature*, **321**, 774–776.
160. Hopfield,J.F., Tank,D.W., Greengard,P. and Huganir,R.L. (1988) Functional modulation of the nicotinic acetylcholine receptor by tyrosine phosphorylation. *Nature*, **336**, 677–680.
161. Albuquerque,E.X., Deshpande,S.S., Aracava,Y., Alkonodon,M. and Daly,J. (1986) A possible involvment of cyclic AMP in the expression of desensitization of the nicotinic acetylcholine receptor. *FEBS Lett.*, **199**, 113–120.
162. Middleton,P., Jaramillo,F. and Schuetze,S.M. (1986) Forskolin increases the rate of acetylcholine receptor desensitization at rat soleus endplates. *Proc. Natl. Acad. Sci. USA*, **83**, 4967–4971.
163. Mulle,C., Benoit,P., Pinset,C., Roa,M. and Changeux,J.-P. (1988) Calcitonin gene-related peptide enhances the rate of desensitization of the nicotinic acetylcholine receptor in cultured mouse muscle cells. *Proc. Natl. Acad. Sci. USA*, **85**, 5728–5732.
164. Eusebi,F., Molinaro,M. and Zani,B.M. (1985) Agents that activate protein kinase C reduce acetylcholine sensitivity in cultured myotubes. *J. Cell Biol.*, **100**, 1339–1342.
165. Betz,H. and Changeux,J.-P. (1979) Regulation of muscle acetylcholine receptor synthesis *in vitro* by cyclic nucleotide derivatives. *Nature*, **278**, 749–752.
166. Blosser,J.C. and Appel,S.H. (1980) Regulation of acetylcholine receptor by cyclic AMP. *J. Biol. Chem.*, **255**, 1235–1238.
167. Fontaine,B., Klarsfeld,A. and Changeux,J.-P. (1987) Calcitonin gene-related peptide and muscle activity regulate acetylcholine receptor α-subunit mRNA levels by distinct intracellular pathways. *J. Cell Biol.*, **105**, 1337–1342.
168. Ross,A.F., Rapuano,M., Schmidt,J.H. and Prives,J.M. (1987) Phosphorylation and assembly of nicotinic acetylcholine receptor subunits in cultured chick muscle cells. *J. Biol. Chem.*, **262**, 14640–14647.
169. Green,W.N. and Claudio,T. (1988) Differences in the phosphorylation of unassembled, assembled but cytoplasmic, and surface acetylcholine receptors. *Soc. Neurosci. Abs.*, **14**, 1045 (419.3).
170. Karlin,A. (1969) Chemical modification of the active site of the acetylcholine receptor. *J. Gen. Physiol.*, **54**, 245s–264s.
171. Kyte,J. and Doolittle,R.F. (1982) A simple method for displaying the hydropathic character of a protein. *J. Mol. Biol.*, **157**, 105–132.
172. Kao,P.N. and Karlin,A. (1986) Acetylcholine receptor binding site contains a disulfide cross-link between adjacent half-cystinyl residues. *J. Biol. Chem.*, **261**, 8085–8088.
173. Guy,R.H. (1984) A structural model of the acetylcholine receptor channel based on partition energy and helix packing calculations. *Biophys. J.*, **45**, 249–259.
174. Finer-Moore,J. and Stroud,R.M. (1984) Amphipathic analysis and possible formation of the ion channel in an acetylcholine receptor. *Proc. Natl. Acad. Sci. USA*, **81**, 155–159.
175. Wennogle,L.P. and Changeux,J.-P. (1980) Transmembrane orientation of proteins present in acetylcholine receptor-rich membranes from *Torpedo marmorata* studied by selective proteolysis. *Eur. J. Biochem.*, **106**, 381–393.
176. Ratnam,M. and Lindstrom,J. (1984) Structural features of the nicotinic acetylcholine receptor revealed by antibodies to synthetic peptides. *Biochem. Biophys. Res. Commun.*, **12**, 1225–1233.
177. Lindstrom,J., Criado,M., Hochschwener,S., Fox,J.L. and Sarin,V. (1984) Immunochemical tests of acetylcholine receptor subunit models. *Nature*, **311**, 573–575.
178. Young,E.F., Ralston,E., Blake,J., Ramachandran,J., Hall,Z.H. and Stroud,R.M. (1985) Topological mapping of acetylcholine receptor: evidence for a model with five transmembrane segments and a cytoplasmic COOH-terminal peptide. *Proc. Natl. Acad. Sci. USA*, **82**, 626–630.
179. Criado,M., Hochschwender,S., Sarin,V., Fox,J.L. and Lindstrom,J. (1985) Evidence for unpredicted transmembrane domains in acetylcholine receptor subunits. *Proc. Natl. Acad. Sci. USA*, **82**, 2004–2008.
180. Ratnam,M., Nguyen,C.L., Rivier,J., Sargent,P.B. and Lindstrom,J. (1986) Transmembrane topography of nicotinic acetylcholine receptor: immunological tests

contradict theoretical predictions based on hydrophobicity profiles. *Biochemistry*, **25**, 2633–2643.
181. Kordossi,A.A. and Tzartos,S.J. (1987) Conformation of cytoplasmic segments of acetylcholine receptor α- and β-subunits probed by monoclonal antibodies: sensitivity of the antibody competition approach. *EMBO J.*, **6**, 1605–1610.
182. Neumann,D., Fridkin,M. and Fuchs,S. (1984) Anti-acetylcholine receptor response achieved by immunization with a synthetic peptide from the receptor sequence. *Biochem. Biophys. Res. Commun.*, **121**, 673–679.
183. LaRochelle,W.J., Wray,B.E., Sealock,R. and Froehner,S.C. (1985) Immunochemical demonstration that amino acids 360–377 of the acetylcholine receptor gamma-subunit are cytoplasmic. *J. Cell Biol.*, **100**, 684–691.
184. Mulac-Jericevic,B., Kurisaki,J. and Atassi,M.Z. (1987) Profile of the continuous antigenic regions on the extracellular part of the α chain of an acetylcholine receptor. *Proc. Natl. Acad. Sci. USA*, **84**, 3633–3637.
185. Toyoshima,C. and Unwin,P.N.T. (1988) Ion channel of acetylcholine receptor reconstructed from images of postsynaptic membranes. *Nature*, **336**, 247–250.
186. LaPolla,R.J., Mixter-Mayne,K. and Davidson,N. (1984) Isolation and characterization of a cDNA clone for the complete protein coding region of the δ subunit of the mouse acetylcholine receptor. *Proc. Natl. Acad. Sci. USA*, **81**, 7970–7974.
187. Kubo,T., Noda,M., Takai,T., Tanabe,T., Kayano,T., Shimizu,S., Tanaka,K., Takahashi,H., Hirose,T., Inayama,S., Kikuno,R., Miyata,T. and Numa,S. (1985) Primary structure of δ subunit precursor of calf muscle acetylcholine receptor deduced from cDNA sequence. *Eur. J. Biochem.*, **149**, 5–13.
188. McCrea,P., Popot,J.-L. and Engelman,D. (1986) Accessibility of the acetylcholine receptor δ chain C terminus to hydrophilic reagents in reconstituted vesicles. *Biophys. J.*, **49**, 355a (T-Pos 308).
189. Dunn,S.M.J., Conti-Tronconi,B.M. and Raftery,M.A. (1986) Acetylcholine receptor dimers are stabilized by extracellular disulfide bonding. *Biochem. Biophys. Res. Commun.*, **139**, 830–837.
190. McCrea,P., Popot,J.-L. and Engelman,D.M. (1987) Transmembrane topography of nicotinic acetylcholine receptor δ subunit. *EMBO J.*, **6**, 3619–3626.
191. Wennogle,L.P., Oswald,R., Saitoh,T. and Changeux,J.-P. (1981) Dissection of the 66 000-dalton subunit of the acetylcholine receptor. *Biochemistry*, **20**, 2492–2497.
192. DiPaola,M., Czajkowski,C., Bodkin,M. and Karlin,A. (1988) The C-terminus of the *Torpedo* acetylcholine receptor δ subunit is extracellular. *Soc. Neurosci. Abs.*, **14**, 640 (260.5).
193. Noda,M., Furutani,Y., Takahashi,H., Toyosato,M., Tanabe,T., Shimizu,S., Kikyotani,S., Kayano,T., Hirose,T., Inayama,S. and Numa,S. (1983) Cloning and sequence analysis of calf cDNA and human genomic DNA encoding α-subunit precursor of muscle acetylcholine receptor. *Nature*, **305**, 818–823.
194. Tanabe,T., Noda,M., Furutani,Y., Takai,T., Takahashi,H., Tanaka,K., Hirose,T., Inayama,S. and Numa,S. (1984) Primary structure of β subunit precursor of calf muscle acetylcholine receptor deduced from cDNA sequence. *Eur. J. Biochem.*, **144**, 11–17.
195. Schubert,D., Harris,A.J., Devine,C.E. and Heinemann,S. (1974) Characterization of a unique muscle cell line. *J. Cell Biol.*, **61**, 398–413.
196. Boulter,J., Luyten,W., Evans,K., Mason,P., Ballivet,M., Goldman,D., Stengelin,S., Martin,G., Heinemann,S. and Patrick,J. (1985) Isolation of a clone coding for the α subunit of a mouse acetylcholine receptor. *J. Neurosci.*, **5**, 2545–2552.
197. Isenberg,K.E., Mudd,J., Shah,V. and Merlie,J.P. (1986) Nucleotide sequence of the mouse muscle nicotinic acetylcholine receptor α subunit. *Nucleic Acids Res.*, **14**, 5111.
198. Buonanno,A., Mudd,J., Shah,V. and Merlie,J.P. (1986) A universal oligonucleotide probe for acetylcholine receptor genes. *J. Biol. Chem.*, **261**, 16451–16458.
199. Patrick,J., Boulter,J., Goldman,D., Gardner,P. and Heinemann,S. (1987) Molecular biology of nicotinic acetylcholine receptors. *Ann. NY Acad. Sci.*, **505**, 194–207.
200. Hartman,D.S. and Claudio,T. (1988) Isolation of a second muscle AChR alpha-like subunit from *Xenopus laevis*. *Soc. Neurosci. Abs.*, **14**, 1045 (419.5).
201. Kistler,J., Stroud,R.M., Klymkowsky,M.W., Lalancette,R.A. and Fairclough,R.H. (1982) Structure and function of an acetylcholine receptor. *Biophys. J.*, **37**, 371–383.

202. Huang,L.M., Catterall,W.A. and Ehrenstein,G. (1978) Selectivity of cations and nonelectrolytes for acetylcholine-activated channels in cultured muscle cells. *J. Gen. Physiol.*, **71**, 397–410.
203. Fairclough,R.H., Miake-Lye,R.C., Stroud,R.M., Hodgson,K.O. and Doniach,S. (1986) Location of terbium binding sites on acetylcholine receptor-enriched membranes. *J. Mol. Biol.*, **189**, 673–680.
204. Dani,J.A. and Eisenman,G. (1987) Monovalent and divalent cation permeation in acetylcholine receptor channels. *J. Gen.Physiol.*, **89**, 959–983.
205. Dani,J. (1986) Ion-channel entrances influence permeation. *Biophys. J.*, **49**, 607–618.
206. Eisenman,G. and Dani,J.A. (1987) An introduction to molecular architecture and permeability of ionic channels. *Annu. Rev. Biophys. Biophys. Chem.*, **16**, 205–226.
207. Imoto,K., Methfessel,C., Sakmann,B., Mishina,M., Mori,Y., Konno,T., Fukuda,K., Kuraske,M., Bujo,H., Fujita,Y. and Numa,S. (1986) Location of a δ-subunit region determining ion transport through the acetylcholine receptor channel. *Nature*, **324**, 670–674.
208. Imoto,K., Busch,C., Sakmann,B., Mishina,M., Konno,T., Nakai,J., Bujo,H., Mori,Y., Fukuda,K. and Numa,S. (1988) Rings of negatively charged amino acids determine the acetylcholine receptor channel conductance. *Nature*, **335**, 645–648.
209. Hopp,T.P. and Woods,K.R. (1981) Prediction of protein antigenic determinants from amino acid sequences. *Proc. Natl. Acad. Sci. USA*, **78**, 3824–3828.
210. Engelman,D.M., Steitz,T.A. and Goldman,A. (1986) Identifying nonpolar transbilayer helices in amino acid sequences of membrane proteins. *Annu. Rev. Biophys. Biophys. Chem.*, **15**, 321–353.
211. Karlin,A., DiPaola,M., Kao,P.N. and Lobel,P.N. (1987) In *Proteins of Excitable Membranes*. B.Hille and D.M.Fambrough (eds), John Wiley and Sons, New York, pp. 43–65.
212. Carp,J.S., Aronstam,R.S., Witkop,B. and Albuquerque,E.X. (1983) Electrophysiological and biochemical studies on enhancement of desensitization by phenothiazine neuroleptics. *Proc. Natl. Acad. Sci. USA*, **80**, 310–314.
213. Adams,P.R. and Feltz,A. (1980) Quinacrine (mepacrine) action at frog end-plate. *J. Physiol.*, **306**, 261–281.
214. Cox,R.N., Kaldany,R.-R., DiPaola,M. and Karlin,A. (1985) Time-resolved photolabeling by quinacrine azide of a noncompetitive inhibitor site of the nicotinic acetylcholine receptor in a transient, agonist-induced state. *J. Biol. Chem.*, **260**, 7186–7193.
215. Heidmann,T., Oswald,R. and Changeux,J.-P. (1983) Multiple sites of action or noncompetitive blockers of acetylcholine receptor rich membrane fragments from *Torpedo marmorata*. *Biochemistry*, **22**, 3112–3127.
216. Tobimatsu,T., Fujita,Y., Fukuda,K., Tanaka,K.-i., Mori,Y., Konno,T., Mishina,M. and Numa,S. (1987) Effects of substitution of putative transmembrane segments on nicotinic acetylcholine receptor function. *FEBS Letts.*, **222**, 56–62.
217. Takai,T., Noda,M., Mishina,M., Shimizu,S., Furutani,Y., Kayano,T., Ikeda,T., Kubo,T., Takahashi,H., Takahashi,T., Kuno,M. and Numa,S. (1985) Cloning, sequencing and expression of cDNA for a novel subunit of acetylcholine receptor from calf muscle. *Nature*, **315**, 761–764.
218. Mishina,M., Takai,T., Imoto,K., Noda,M., Takahashi,T., Numa,S., Methfessel,C. and Sakmann,B. (1986) Molecular distinction between fetal and adult forms of muscle acetylcholine receptor. *Nature*, **321**, 406–411.
219. Witzemann,V., Barg,B., Nishikawa,Y., Sakmann,B. and Numa,S. (1987) Differential regulation of muscle acetylcholine receptor γ- and ε-subunit mRNA. *FEBS Letts.*, **223**, 104–112.
220. Bluescript DNA sequencing system from Stratagene, 3770 Tansy Street, San Diego, CA 92121.
221. Palazzolo,M.J. and Meyerowitz,E.M. (1987) A family of lambda phage cDNA cloning vectors, λSWAJ, allowing amplification of RNA sequences. *Gene*, **52**, 197–206.
222. LambdaGEM vectors from Promega, 2880 S. Fish Hatchery Road, Madison, WI 5371–5305.
223. Lentz,T.L. and Wilson,P.T. (1988) Neurotoxin-binding site on the acetylcholine receptor. *Internat. Rev. Neurobiol.*, **29**, 117–160.
224. Barkas,T., Mauron,A., Roth,B., Alliod,C., Tzartos,S.J. and Ballivet,M. (1987)

Mapping the main immunogenic region and toxin-binding site of the nicotinic acetylcholine receptor. *Science,* **235,** 77–80.
225. Gershoni,J.M. (1987) Expression of the α-bungarotoxin binding site of the nicotinic acetylcholine receptor by *Escherichia coli* transformants. *Proc. Natl. Acad. Sci. USA,* **84,** 4318–4321.
226. White,M.M., Mixter-Mayne,K., Lester,H.A. and Davidson,N. (1985) Mouse – *Torpedo* hybrid acetylcholine receptors: functional homology does not equal sequence homology. *Proc. Natl. Acad. Sci. USA,* **82,** 4852–4856.
227. Melton,D.A., Krieg,P.A., Rebagliati,M.R., Maniatis,T., Zinn,K. and Green,M.R. (1984) Efficient *in vitro* synthesis of biologically active RNA and RNA hybridization probes from plasmids containing a bacteriophage SP6 promoter. *Nucleic Acids Res.,* **12,** 7035–7056.
228. Methfesel,C., Witzemann,V., Takahashi,T., Numa,S. and Sakmann,B. (1986) Patch clamp measurements on *Xenopus laevis* oocytes: currents through endogenous channels and implanted acetylcholine receptor and sodium channels. *Pfluger's Arch.,* **407,** 577–588.
229. Botstein,D. and Davis,R.W. (1982) Principles and practice of recombinant DNA research with yeast. In *Molecular Biology of the Yeast Saccharomyces: Metabolism and Gene Expression.* J.N.Strathern, E.W.Jones and J.R.Broach (eds), Cold Spring Harbor Laboratory Press, New York, pp. 607–636.
230. Broach,J.R., Li,Y.-Y., Wu,L.-C.C. and Jayaram,M. (1983) Vectors for high-level, inducible expression of cloned genes in yeast. In *Experimental Manipulation of Gene Expression.* M.Inouye (ed.), Academic Press, Inc., Orlando, Florida, pp. 83–117.
231. Fujita,N., Nelson,N., Fox,T.D., Claudio,T., Lindstrom,J., Riezman,H. and Hess,G.P. (1986) Biosynthesis of the *Torpedo californica* acetylcholine receptor α subunit in yeast. *Science,* **231,** 1284–1287.
232. Fujita,N., Sweet,M.T., Fox,T.D., Nelson,N., Claudio,T., Lindstrom,J.M. and Hess,G.P. (1986) Expression of cDNAs for acetylcholine receptor subunits in the yeast cell plasma membrane. *Biochem. Soc. Symp.,* **52,** 41–56.
233. Sweet,M.T., Lindstrom,J., Fujita,N., Jansen,K., Min,C.K., Claudio,T., Nelson,N., Fox,T.D. and Hess,G.P. (1988) Expression of acetylcholine receptor subunits in *Saccharomyces cerevisiae* (yeast). *Curr. Top. Membr. Transp.,* **33,** 197–218.
234. Jansen,K.U., Conroy,W.G., Claudio,T., Fox, T.D., Fujita,N., Hamill,O., Lindstrom,J., Luther,M., Nelson,N., Ryan,K.A., Sweet,M.T. and Hess,G.P. (1989) Expression of the four subunits of *Torpedo californica* nicotinic acetylcholine receptor in *Saccharomyces cerevisiae. J. Biol. Chem.,* **264,** in press.
235. Ballou,C.E. (1982) Yeast cell wall and cell surface. In *Molecular Biology of the Yeast Saccharomyces: Metabolism and Gene Expression.* J.N.Strathern, E.W.Jones and J.R.Broach (eds), Cold Spring Harbor Laboratory Press, New York, pp. 335–360.
236. Kukuruzinska,M.A., Bergh,M.L.E. and Jackson,B.J. (1987) Protein glycosylation in yeast. *Annu. Rev. Biochem.,* **56,** 915–944.
237. Schekman,R. and Novick,P. (1982) The secretory process and yeast cell surface's assembly. In *Molecular Biology of Yeast Saccharomyces: Metabolism and Gene Expression.* J.N.Strathern, E.W.Jones and J.R.Broach (eds), Cold Spring Harbor Laboratory Press, New York, pp. 361–398.
238. Claudio,T., Green,W.N., Hartman,D.S., Hayden,D., Paulson,H.L., Sigworth,F.J., Sine,S.M. and Swedlund,A. (1987) Genetic reconstitution of functional acetylcholine receptor channels in mouse fibroblasts. *Science,* **238,** 1688–1694.
239. Hartman,D.S., Poo,M.-m., Green,W.N., Ross,A.F. and Claudio,T. (1989) Synaptic contact between embryonic neurons and acetylcholine receptor-fibroblasts. *J. Physiol. Paris,* in press.
240. Sine,S.M., Claudio,T. and Sigworth,F.J. (1988) Functional properties of *Torpedo* acetylcholine receptors genetically reconstituted in mouse fibroblasts. *Biophys. J.,* **53,** 637a (Th-AM-C5).
241. Weiland,G. and Taylor,P. (1979) Ligand specificity of state transitions in the cholinergic receptor: behavior of agonists and antagonists. *Mol. Pharmacol.,* **15,** 197–212.
242. Hess,G.P., Pasquale,E.B., Walker,J.W. and McNamee,M.G. (1982) Comparison of acetylcholine receptor-controlled cation flux in membrane vesicles from *Torpedo californica* and *Electrophorus electricus*: channel kinetic measurements in the

millisecond region. *Proc. Natl. Acad. Sci. USA,* **79**, 963–967.
243. Paulson,H.P. and Claudio,T. (1987) Analysis of *Torpedo californica* nicotinic acetylcholine receptor subunits expressed in mammalian fibroblasts and muscle cell lines. *J. Cell Biol.,* **105**, 62a (341).
244. Paulson,H.L. and Claudio,T. (1988) Temperature-sensitive expression of all-*Torpedo* and *Torpedo*–rat hybrid AChRs in mammalian cells. *Soc. Neurosci. Abs.,* **14**, 1045 (419.4).
245. Claudio,T., Hartman,D.S., Green,W.N., Ross,A.F., Paulson,H.L. and Hayden,D. (1989) Stable expression of multisubunit protein complexes in mammalian cells. In *NATO ASI Series H, Cell Biology, Molecular Biology of Neuroreceptors and Ion Channels.* A.Maelicke (ed.), Springer-Verlag, Berlin, Vol. 32, pp. 469–480.
246. Mishina,M., Kurosaki,T., Tobimatsu,T., Morimoto,Y., Noda,M., Yamamoto,T., Terao,M., Lindstrom,J., Takahashi,T., Kuno,M. and Numa,S. (1984) Expression of functional acetylcholine receptor from cloned cDNAs. *Nature,* **307**, 604–608.
247. Kurosaki,T., Fukuda,K., Konno,T., Mori,Y., Tanaka,K., Mishina,M. and Numa,S. (1987) Functional properties of nicotinic acetylcholine receptor subunits expressed in various combinations. *FEBS Letts.,* **214**, 253–258.
248. Haggerty,J.G. and Froehner,S.C. (1981) Restoration of ^{125}I-α-bungarotoxin binding activity to the α subunit of *Torpedo* acetylcholine receptor isolated by gel electrophoresis in sodium dodecyl sulfate. *J. Biol. Chem.,* **256**, 8294–8297.
249. Gershoni,J.M., Hawrot,E. and Lentz,T.L. (1983) Binding of α-bungarotoxin to isolated α subunit of the acetylcholine receptor of *Torpedo californica*: quantitative analysis with protein blots. *Proc. Natl. Acad. Sci. USA,* **80**, 4973–4977.
250. White,M.M. (1987) Deltaless acetylcholine receptors are highly voltage-dependent. *Soc. Neurosci. Abs.,* **13**, 798 (223.12).
251. Mixter-Mayne,K., Yoshii,K., Yu,L., Lester,H.A. and Davidson,N. (1987) Expression of mouse–*Torpedo* acetylcholine receptor subunit chimeras and hybrids in *Xenopus* oocytes. *Mol. Brain Res.,* **2**, 191–197.
252. Yoshii,K., Yu,L., Mixter-Mayne,K., Davidson,N. and Lester,H.A. (1987) Equilibrium properties of mouse–*Torpedo* acetylcholine receptor hybrids expressed in *Xenopus* oocytes. *J. Gen. Physiol.,* **90**, 553–573.
253. Leonard,R.J., Yu,L., Labarca,C., Davidson,N. and Lester,H.A. (1987) Effects of a local anesthetic (QX-222) on gating properties of mouse–*Torpedo* hybrid acetylcholine receptors expressed in *Xenopus* oocytes. *Soc. Neurosci. Abs.,* **13**, 97 (30.9).
254. Yu,L., Leonard,R.J., Labarca,C., Davidson,N. and Lester,H.A. (1987) Channel duration mainly determines voltage sensitivity in mouse–*Torpedo* acetylcholine receptor hybrids. *Soc. Neurosci. Abs.,* **13**, 97 (30.10).
255. Sakmann,B., Methfessel,C., Mishina,M., Takahashi,T., Takai,M., Kurosaki,K., Fukuda,K. and Numa,S. (1985) Role of acetylcholine receptor subunits in gating of the channel. *Nature,* **318**, 538–543.
256. Claudio,T. (1986) Establishing a system for the stable expression of *Torpedo* acetylcholine receptors. In *NATO ASI Series H: Cell Biology, Vol. 3, Nicotinic Acetylcholine Receptor.* A.Maelicke (ed.), Springer-Verlag, Berlin, pp. 431–435.
257. Claudio,T., Paulson,H.L., Hartman,D., Sine,S. and Sigworth,F.J. (1988) Establishing a stable expression system for studies of acetylcholine receptors. *Curr. Top. Membr. Transp.,* **33**, 219–247.
258. Blount,P. and Merlie,J.P. (1988) Native folding of an acetylcholine receptor α subunit expressed in the absence of other receptor subunits. *J. Biol. Chem.,* **263**, 1072–1080.
259. Yaffe,D. (1968) Retention of differentiation potentialities during prolonged cultivation of myogenic cells. *Proc. Natl. Acad. Sci. USA,* **82**, 4852–4856.
260. Neubig,R.R. and Cohen,J.B. (1979) Equilibrium binding of [^{3}H]tubocurarine and [^{3}H]acetylcholine by *Torpedo* postsynaptic membranes: stoichiometry and ligand interactions. *Biochemistry,* **18**, 5464–5475.
261. Merlie,J.P., Sebbane,R., Gardner,S., Olson,E. and Lindstrom,J. (1983) The regulation of acetylcholine receptor expression in mammalian muscle. Cold Spring Harbor Symp. Quant. Biol., **48**, 135–146.
262. Klarsfeld,A., Devillers-Thiery,A., Giraudat,J. and Changeux,J.-P. (1984) A single gene codes for the nicotinic acetylcholine receptor α-subunit in *Torpedo marmorata*: structural and developmental implications. *EMBO J.,* **3**, 35–41.
263. Schuetze,S.M., Vicini,S. and Hall,Z.W. (1985) Myasthenic serum selectively blocks

acetylcholine receptors with long channel open times at developing rat endplates. *Proc. Natl. Acad. Sci. USA,* **82**, 2533–2537.
264. Pedersen,S.E., Dreyer,E.B. and Cohen,J.B. (1986) Location of ligand-binding sites on the nicotinic acetylcholine receptor α-subunit. *J. Biol. Chem.,* **261**, 13735–13743.
265. Wise,D.S., Wall,J. and Karlin,A. (1981) Relative locations of the β and δ chains of the acetylcholine receptor determined by electron microscopy of isolated receptor trimer. *J. Biol. Chem.,* **256**, 12624–12627.
266. Holtzman,E., Wise,D., Wall,J. and Karlin,A. (1982) Electron microscopy of complexes of isolated acetylcholine receptor, biotinyl-toxin, and avidin. *Proc. Natl. Acad. Sci. USA,* **79**, 310–314.
267. Zingsheim,H.P., Barrantes,F.J., Frank,J., Hanicke,W. and Neugebauer,D.C. (1982) Direct structural localization of two toxin-recognition sites on an ACh receptor protein. *Nature,* **299**, 81–84.
268. Karlin,A., Holtzman,E., Yodh,N., Lobel,P., Wall,J. and Hainfeld,J. (1987) The arrangement of the subunits of the acetylcholine receptor of *Torpedo californica*. *J. Biol. Chem.,* **258**, 6678–6681.
269. Hamilton,S.L., Pratt,D.R. and Eaton,D.C. (1985) Arrangement of the subunits of the nicotinic acetylcholine receptor of *Torpedo californica* as determined by α-neurotoxin cross-linking. *Biochemistry,* **24**, 2210–2219.
270. Fairclough,R.H., Finer-Moore,J., Love,R.A., Kristofferson,D., Desmeules,P.J. and Stroud,R.M. (1983) Subunit organization and structure of an acetylcholine receptor. *Cold Spring Harbor Symp. Quant. Biol.,* **48**, 9–20.
271. Kubalek,E., Ralston,S., Lindstrom,J. and Unwin,N. (1987) Location of subunits within the acetylcholine receptor by electron image analysis of tubular crystals from *Torpedo marmorata*. *J. Cell Biol.,* **105**, 9–18.
272. Tzartos,S.J., Rand,D.E., Einarson,B.L. and Lindstrom,J. (1981) Mapping of surface structures of *Electrophorus* acetylcholine receptor using monoclonal antibodies. *J. Biol. Chem.,* **256**, 8635–8645.
273. Tzartos,S.J., Langeberg,L., Hochschwender,S., Swanson,L.W. and Lindstrom,J. (1981) Characteristics of monoclonal antibodies to denatured *Torpedo* and to native calf acetylcholine receptors: species, subunit and region specificity. *J. Neuroimmunol.,* **10**, 235–253.
274. Karlin,A. (1987) Going around in receptor circles. *Nature,* **329**, 286–287.
275. Merlie,J.P. and Lindstrom,J. (1983) Assembly *in vivo* of mouse muscle acetylcholine receptor: identification of an α subunit species that may be an assembly intermediate. *Cell,* **34**, 747–757.
276. Numa,S., Takahashi,H., Tanabe,T., Toyosato,M., Furutani,Y. and Kiyotani,S. (1983) Molecular structure of the nicotinic acetylcholine receptor. *Cold Spring Harbor Symp. Quant. Biol.,* **48**, 57–69.
277. Hershey,N.D., Noonan,D.J., Mixter,K.S., Claudio,T. and Davidson,N. (1983) Structure and expression of genomic clones coding for the δ-subunit of the *Torpedo* acetylcholine receptor. *Cold Spring Harbor Symp. Quant. Biol.,* **48**, 79–82.
278. Ballivet,M., Nef,P., Stalder,R. and Fulpius,B. (1983) Genomic sequences encoding the α-subunit of acetylcholine receptor are conserved in evolution. *Cold Spring Harbor Symp. Quant. Biol.,* **48**, 83–87.
279. Nef,P., Oneyser,C., Barkas,T. and Ballivet,M. (1986) Acetylcholine receptor related genes expressed in the nervous system. In *NATO ASI Series H: Cell Biology, Vol. 3, Nicotinic Acetylcholine Receptor.* A.Maelicke (ed.), Springer-Verlag, Berlin, pp. 417–422.
280. Klarsfeld,A., Daubas,P., Bourachot,B. and Changeux,J.-P. (1987) A 5'-flanking region of the chicken acetylcholine receptor α-subunit gene confers tissue specificity and developmental control of expression in transfected cells. *Mol. Cell. Biol.,* **7**, 951–955.
281. Gardner,P.D., Heinemann,S. and Patrick,J. (1987) Deletion analysis of the nicotinic acetylcholine receptor γ-subunit gene promoter. *Soc. Neurosci. Abs.,* **13**, 1287 (357.18).
282. Crowder,C.M. and Merlie,J.P. (1986) DNase I-hypersensitive sites surround the mouse acetylcholine receptor δ-subunit gene. *Proc. Natl. Acad. Sci. USA,* **83**, 8405–8409.
283. Evans,S.M., Gardner,P.D., Heinemann,S. and Patrick,J. (1987) Identification of transcriptional regulatory elements of the muscle nicotinic acetylcholine receptor δ-subunit gene. *Soc. Neurosci. Abs.,* **13**, 1287 (357.18).

284. Heidmann,O., Buonanno,A., Geoffroy,B., Robert,B., Guenet,J.-L., Merlie,J.P. and Changeux,J.-P. (1986) Chromosomal localization of muscle nicotinic acetylcholine receptor genes in the mouse. *Science,* **234**, 866–868.
285. Fambrough,D.M. (1970) Acetylcholine sensitivity of muscle fiber membranes: mechanism of regulation by motor neurons. *Science,* **168**, 372–373.
286. Merlie,J.P., Isenberg,K.E., Russell,S.D. and Sanes,J.R. (1984) Denervation supersensitivity in skeletal muscle: analysis with a cloned cDNA probe. *J. Cell Biol.,* **99**, 332–335.
287. Goldman,D., Boulter,J., Heinemann,S. and Patrick,J. (1985) Muscle innervation increases the levels of two mRNAs coding for the acetylcholine receptor α-subunit. *J. Neurosci.,* **5**, 2553–2558.
288. Klarsfeld,A. and Changeux,J.-P. (1985) Activity regulates the levels of acetylcholine receptor α-subunit mRNA in cultured chicken myotubes. *Proc. Natl. Acad. Sci. USA,* **82**, 4558–4562.
289. Shieh,B.H., Ballivet,M. and Schmidt,J. (1987) Quantitation of an alpha subunit splicing intermediate: evidence for transcriptional activation in the control of acetylcholine receptor expression in denervated chick skeletal muscle. *J. Cell Biol.,* **104**, 1337–1341.
290. Shieh,B.H., Ballivet,M. and Schmidt,J. (1988) Acetylcholine receptor synthesis rate and levels of receptor subunit messenger RNAs in chick muscle. *Neuroscience,* **24**, 175–187.
291. Covault,J., Merlie,J.P., Goridis,C. and Sanes,J.R. (1986) Molecular forms of N-CAM and its RNA in developing and denervated skeletal muscle. *J. Cell Biol.,* **102**, 731–739.
292. Buonanno,A. and Merlie,J.P. (1986) Transcriptional regulation of nicotinic acetylcholine receptor genes during muscle development. *J. Biol. Chem.,* **261**, 11452–11455.
293. Goldman,D., Brenner,H. and Heinemann,S. (1988) Acetylcholine receptor α-, β-, γ- and δ-subunit mRNA levels are regulated by muscle activity. *Neuron,* **1**, 329–333.
294. Chow,I. and Cohen,M.W. (1983) Developmental changes in the distribution of acetylcholine receptors in the myotomes of *Xenopus laevis*. *J. Gen. Physiol.,* **339**, 553–571.
295. Brehm,P., Kidokoro,Y. and Moody-Corbett,F. (1984) Acetylcholine receptor channel properties during development of *Xenopus* muscle cells in culture. *J. Gen. Physiol.,* **357**, 203–217.
296. Hermans-Borgmeyer,I., Zopf,D., Ryseck,R., Hovemann,B., Betz,H. and Gundelfinger,E.D. (1986) Primary structure of a developmentally regulated nicotinic acetylcholine receptor protein from *Drosophila*. *EMBO J.,* **5**, 1503–1508.
297. Wada,K., Ballivet,M., Boulter,J., Connolly,J., Deneris,E., Wada,E., Swanson,L., Heinemann,S. and Patrick,J. (1987) Isolation and characterization of a gene coding for a rat brain nicotinic acetylcholine receptor α-subunit. *Soc. Neurosci. Abs.,* **13**, 940 (260.12).
298. Wada,K., Ballivet,M., Boulter,J., Connolly,J., Wada,E., Deneris,E., Swanson,L., Heinemann,S. and Patrick,J. (1988) Functional expression from cDNA clones of a new pharmacological subtype of brain nicotinic acetylcholine receptor. *Science,* **240**, 330–334.
299. Goldman,D.J., Deneris,E., Luyten,W., Kochhar,S., Patrick,J. and Heinemann,S. (1987) Members of a nicotinic acetylcholine receptor gene family are expressed in different regions of the mammalian central nervous system. *Cell,* **48**, 965–973.
300. Boulter,J., Connolly,J., Deneris,E., Goldman,D., Heinemann,S. and Patrick,J. (1987) Functional expression of two neuronal nicotinic acetylcholine receptors from cDNA clones identifies a gene family. *Proc. Natl. Acad. Sci. USA,* **84**, 7763–7767.
301. Deneris,E., Connolly,J., Boulter,J., Wada,E., Wada,K., Swanson,L., Patrick,J. and Heinemann,S. (1988) Primary structure and expression of β2: a novel subunit of neuronal nicotinic acetylcholine receptors. *Neuron,* **1**, 45–54.
302. Nef,P., Oneyser,C., Alliod,C., Couturier,S. and Ballivet,M. (1988) Genes expressed in the brain define three distinct neuronal nicotinic acetylcholine receptors. *EMBO J.,* **7**, 595–601.
303. Bossy,B., Ballivet,M. and Spierer,P. (1988) Conservation of neuronal nicotinic acetylcholine receptors from *Drosophila* to vertebrate central nervous systems. *EMBO J.,* **7**, 611–618.

304. Boyd,R.T., Jacob,M.H., Couturier,S., Ballivet,M. and Berg,D.K. (1987) Expression and regulation of neuronal acetylcholine receptor mRNA in chick ciliary ganglia. *Neuron,* **1**, 495–502.
305. Breer,H., Hinz,G., Madler,U. and Hanke,W. (1986) Identification and reconstitution of a neuronal acetylcholine receptor from insects. In *NATO ASI Series H: Cell Biology, Volume 3, Nicotinic Acetylcholine Receptor.* A.Maelicke (ed.), Springer-Verlag, Berlin, pp. 319–332.
306. Hanke,W. and Breer,H. (1986) Channel properties of an insect neuronal acetylcholine receptor protein reconstituted in planar lipid bilayers. *Nature,* **321**, 171–174.
307. Hanke,W. and Breer,H. (1987) Characterization of the channel properties of a neuronal acetylcholine receptor reconstituted into planar lipid bilayers. *J. Gen. Physiol.,* **90**, 855–879.
308. Schmitt,B., Knaus,P., Becker,C.-M. and Betz,H. (1987) The M_r 93 000 polypeptide of the postsynaptic glycine receptor complex is a peripheral membrane protein. *Biochemistry,* **26**, 805–811.
309. Pritchett,D., Schofield,P., Sontheimer,H., Ymer,S., Kettenmann,H. and Seeburg,P. (1988) GABA receptor cDNAs expressed in transfected cells and studied by patch-clamp and binding assay. *Soc. Neurosci. Abs.,* **14**, 641 (260.16).
310. Levitan,E.S., Blair,L.A.C., Dionne,V.E. and Barnard,E.A. (1988) Biophysical and pharmacological properties of $GABA_A$ receptor subunit clones expressed in oocytes. *Soc. Neurosci. Abs.,* **14**, 1045 (419.8).
311. Blair,L.A.C., Levitan,E.S., Marshall,J., Dionne,V.E. and Barnard,E.A. (1988) Single subunits of the $GABA_A$ receptor form ion channels with properties of the native receptor. *Science,* **242**, 577–579.
312. Pritchett,D., Sontheimer,H., Gorman,C.M., Kettenmann,H., Seeburg,P. and Schofield,P. (1988) Transient expression shows ligand gating and allosteric potentiation of $GABA_A$ receptor subunits. *Science,* **242**, 1306–1308.
313. Agnew,W.S., Levinson,S.R., Brabson,J.S. and Raftery,M.A. (1978) Purification of the tetrodotoxin-binding component associated with the voltage-sensitive sodium channel from *Electrophorus electricus* electroplax membranes. *Proc. Natl. Acad. Sci. USA,* **75**, 2606–2610.
314. Miller,J.A., Agnew,W.S. and Levinson,.SR. (1984) Principal glycopeptide of the tetrodotoxin/saxitoxin binding protein from *Electrophorus electricus*: isolation and partial chemical and physical characterization. *Biochemisry,* **22**, 462–470.
315. Lombet,A. and Lazdunski,M. (1984) Characterization, solubilization, affinity labeling and purification of the cardiac Na^+ channel using *Tityus* toxin gamma. *Eur. J. Biochem.,* **141**, 651–660.
316. Hartshorne,R.P. and Catterall,W.A. (1984) The sodium channel from rat brain; purification and subunit composition. *J. Biol. Chem.,* **259**, 1667–1675.
317. Barchi,R.L. (1983) Protein components of the purified sodium channel from rat skeletal muscle sarcolemma. *Neurochemistry,* **40**, 1377–1385.
318. Guy,H.R. and Seetharamulu,P. (1986) Molecular model of the action potential sodium channel. *Proc. Natl. Acad. Sci. USA,* **83**, 508–512.
319. Armstrong,C.M. (1981) Sodium channel and gating currents. *Physiol. Rev.,* **61**, 644–683.
320. Kyano,T., Noda,M., Flockerzi,V., Takahashi,H. and Numa,S. (1988) Primary structure of rat brain sodium channel III deduced from the cDNA sequence. *FEBS Letts.,* **228**, 187–194.
321. Auld,V.J., Goldin,A.L., Krafte,D.S., Marshall,J., Dunn,J.M., Catterall,W.A., Lester,H.A., Davidson,N. and Dunn,R.J. (1988) A rat brain Na^+ channel α subunit with novel gating properties. *Neuron,* **1**, 449–461.
322. Noda,M., Ikeda,T., Suzuki,H., Takashima,H., Takahashi,T., Kuno,M. and Numa,S. (1986) Expression of functional sodium channels from cloned cDNA. *Nature,* **322**, 826–828.
323. Stuhmer,W., Methfessel,C., Sakmann,B., Noda,M. and Numa,S. (1987) Patch clamp characterization of sodium channels expressed from rat brain cDNA. *Eur. Biophys. J.,* **14**, 131–138.
324. Suzuki,H., Beckh,S., Kubo,H., Yahagi,N., Ishida,H., Kayano,T., Noda,M. and Numa,S. (1988) Functional expression of cloned cDNA encoding sodium channel III. *FEBS Letts.,* **228**, 195–200.

325. Tanabe,T., Beam,K.G., Powell,J.A. and Numa,S. (1988) Restoration of excitation–contraction coupling and slow calcium current in dysgenic muscle by dihydropyridine receptor complementary DNA. *Nature,* **336**, 134–139.
326. Huynh,T.V., Young,R.A. and Davis,R.W. (1985) Constructing and screening cDNA libraries in λgt10 and λgt11. In *DNA Cloning, Volume 1: A Practical Approach.* D.M.Glover and B.D.Hames (eds), IRL Press, Oxford, pp. 48–78.
327. Chao,M.V., Bothwell,M.A., Ross,A.H., Koprowski,H., Lanahan,A.A., Buck,C.R. and Sehgal,A. (1986) Gene transfer and molecular cloning of the human NGF receptor. *Science,* **232**, 518–521.
328. Radeke,M.J., Misko,T.P., Hsu,C., Herzenberg,L.A. and Shooter,E.M. (1987) Gene transfer and molecular cloning of the rat nerve growth factor receptor. *Nature,* **325**, 593–597.
329. Seed,B. and Aruffo,A. (1987) Molecular cloning of the CD2 antigen, the T-cell erythrocyte receptor, by a rapid immunoselection procedure. *Proc. Natl. Acad. Sci. USA,* **84**, 3365–3369.
330. Lubbert,H., Hoffman,B.J., Snutch,T.P., Levine,A.J., Hartig,P.R., Lester,H.A. and Davidson,N. (1987) cDNA cloning of a serotonin 5-HT$_{1C}$ receptor by electrophysiological assays of mRNA-injected *Xenopus* oocytes. *Proc. Natl. Acad. Sci. USA,* **84**, 4332–4336.

4

Molecular biology of cell adhesion in neural development

Jonathan Covault

1. Introduction

The mature nervous system is composed of an amazingly complex yet highly stereotyped network of intercellular communication. To achieve this, the organism utilizes a variety of diverse processes including regulated cell division and death, cell migration, directed axon growth, and the localized induction of synaptic structures. A wealth of observations of normal and experimental neuroembryology during the past century has provided numerous examples of the importance of cell–cell interactions in modifying and regulating these processes. Perhaps not surprisingly, given the difficulties inherent in developing convenient assays for such complex phenomena as directed axon outgrowth and neuron migration, our understanding of the molecules involved in these interactions is still in its infancy. In spite of these difficulties, a variety of *in-vitro* assays have been developed and successfully used to identify molecules which may be central to cell–cell interactions during neural development.

Many cell–cell interactions in neural development require intimate contact between cell surfaces or between cells and extracellular matrix (ECM) surfaces. Cell surface adhesion molecules are undoubtedly key to the regulation and promotion of such contacts. While cell binding is not as conveniently assayed as enzymatic activity or ion channel flux, the use of simple *in-vitro* assays to study cell–cell adhesion dates back to experiments by Wilson (1) in the early part of this century on the reaggregation of mechanically dissociated marine sponge cells to form functional sponges. During the last decade, combined use of immunological methods and quantitative assays of cell binding have produced rapid progress in the isolation of specific molecules involved in neuron–neuron and neuron–glial adhesion. These are listed in *Table 1*. The identification and biochemical characterization of these molecules

Table 1. Neural adhesion molecules

Molecule	Mol. wt	Expressed by	Key features	Putative functions
NCAM	120, 140, 180 kd	Many embryonic cell types, adult neurons, muscle	Homophilic adhesion. May interact with heparan sulfate proteoglycans. Developmentally regulated forms differ in cell-associated domain produced by alternate exon usage. Contains variable amounts of an unusual PSA which can modulate adhesion. Present on both neuron cell body and processes. Immunoglobulin super family	Neuron–neuron, neuron–glia, neurite–muscle and glia–glia cell adhesion. Histogenesis and axon fasciculation
L1/NgCAM	135, 200 kd most prominent, varies between cells	Post-mitotic neurons, some Schwann cells	Homophilic and heterophilic adhesion. High levels on axons low on neuron cell bodies. Major contributor to axon–axon adhesion. Induced by NGF in sensory ganglia neurons. Immunoglobulin superfamily	Axon fasciculation, neurite outgrowth, granule cell migration
MAG	100 kd	Oligodendroglia, myelinating Schwann cells	Concentrated at the periaxonal membrane of myelinating glial cells. Immunoglobulin superfamily	Axon–myelin sheath adhesion, maintenance of 12–15 nm periaxonal spacing

P$_0$	28–30 kd	Myelinating Schwann cells	Transmembrane glycoprotein, comprises 50% of PNS myelin protein. Expression corresponds with the development of PNS myelin. Immunoglobulin superfamily	Adhesion between external (and perhaps internal) leaflets of Schwann cell membrane in compact myelin
N-cadherin	130 kd	Many embryonic cell types, adult neurons, muscle	Calcium-dependent adhesion. Related to cadherins present in other tissue types	Neuron–neuron, neurite–muscle adhesion. Histogenesis
J1/cytotactin	Multiple forms (190, 200, 220 kd) typical in embryo, 160 kd in adult	Variety of cell types, especially glial cells and fibroblasts	Secreted glycoprotein present on surface of CNS glial cells. Binds to neuron cell surface chondroitin sulfate proteoglycan	Neuron–glia adhesion, granule cell migration along Bergmann glial fibers
AMOG	50 kd	Cerebellar astrocytes and oligodendrocytes	Cell surface glycoprotein	Neuron–glial adhesion, granule cell migration along Bergmann glial fibers
Astrotactin	100 kd	Cerebellar granule cells	Cell surface glycoprotein, adhesive qualities not yet assayed. Markedly reduced levels on *weaver* granule cells	Granule cell–Bergmann glia cell adhesion/recognition
HNK-1 epitope		Subset of neurons and glial cells	Sulfated glucuronic acid attached to several adhesion molecules, NCAM, L1/NgCAM, J1/cytotactin and MAG	Antibodies to epitope can inhibit neuron–glia adhesion

Table 1. (cont.)

Molecule	Mol. wt	Expressed by	Key features	Putative functions
Fibronectin	Dimer of similar 220 kd subunits	Serum—hepatocytes, cellular—many cell types especially fibroblasts and PNS glia	Secreted glycoprotein, functionally distinct forms produced by RNA splicing. Multiple functional domains for binding to ECM and cell surface receptors	Neural crest cell migration, axon outgrowth from PNS neurons
Laminin	900 kd complex: A, 440 kd B1, 225 kd B2, 205 kb	Many cell types. Fibroblasts, Schwann cells, CNS glia of frog and fish. Optic tract of higher vertebrates during development	Secreted glycoprotein complex central to basal lamina. Multiple functional domains for binding other ECM molecules and cell surface receptors	Axon outgrowth in PNS, some areas of CNS
S-laminin	190 kd	Muscle and endothelial cells	Basal lamina protein related to laminin with neuron-binding activity. Present in synaptic but not extra-synaptic muscle fiber basal lamina. Also present in perineurial and arterial basal lamina	Potential trigger for nerve-terminal differentiation
Integrin	125/130/155 kd complex	Many cell types including neurons	Chicken cell surface receptor for laminin and fibronectin. Member of cytoadhesin family of adhesion/recognition proteins	Participates in cell binding to laminin and fibronectin

CNS, central nervous system; NGF, nerve growth factor; PNS, peripheral nervous system; PSA, polysialic acid.

will be described in the first half of this chapter followed by consideration of experiments implicating these cell adhesion molecules in specific processes of neural development. In addition to reviewing results, an attempt has been made to provide examples of a variety of experimental approaches, hopefully giving readers exposure of techniques which could have application for their own particular investigations.

2. A diversity of cell adhesion molecules

2.1 NCAM—the first 'neural cell adhesion molecule'

2.1.1 Identification and biochemical characterization

To isolate molecules mediating neural cell adhesion, Edelman's group (2,3) employed an iterative immunological approach. Polyspecific antibodies to the entire retinal cell surface were developed which, when used as monovalent Fab fragments, could inhibit the normal reaggregation of dissociated neural retinal cells. Assuming this antibody effect was due to a masking of relevant cell surface adhesion molecules, this group fractionated cell surface molecules shed by cultured neural retinal cells and isolated three polypeptides which could neutralize the inhibitory effects of their polyspecific antibodies on cell adhesion. A new antiserum was raised using these active fractions as antigens. This second-generation antiserum also inhibited retinal cell adhesion but, unlike the original antibodies, it was relatively monospecific, selectively immunoprecipitating a 140 kd polypeptide from detergent extracts of retinal cells. The three polypeptides purified earlier apparently represented fragments derived from this molecule as pre-incubation of the antibodies with them prevented the immunoprecipitation of the 140 kd cell-associated protein and neutralized the inhibition of retinal cell aggregation. Molecules closely related to this 140 kd retinal antigen are found throughout the nervous system and have subsequently been referred to as the neural cell adhesion molecule or NCAM. Independently, in more general studies of major brain surface components, monoclonal antibodies to chicken (224-1A6, ref. 4), mouse (BSP-2, ref. 5) and polyclonal antibodies to rat (D2, ref. 6) NCAM have been obtained in other laboratories. While NCAM is an appropriate name for an adhesion molecule isolated from neural cells, the antigen has a widespread distribution among embryonic tissues (7), where it is thought to participate with other adhesion molecules to influence the development of multiple tissues. In the adult, NCAM is largely restricted to the surfaces of neurons, glial cells and striated muscle.

NCAM comprises a family of relatively abundant, related glycoproteins (~180, 140, and 120 kd in adult rat and mouse brain) which result from the translation of alternatively spliced mRNAs originating from a single gene (8–12). The larger, two-brain NCAM forms are both transmembrane glycoproteins which differ in the length of their cytoplasmic domain. In

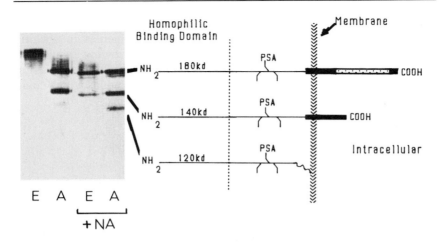

Figure 1. Molecular structure of NCAM. An immunoblot of embryonic (E) and adult (A) rat brain NCAM. Prior to electrophoresis, samples in the last two lanes were treated with neuraminidase (NA) to remove sialic acid. The corresponding linear diagrams of the three forms of NCAM are shown. All three share identical N-terminal sequences, but have different cell-associated C-terminal domains. The stippled box in the 180 kd form represents an additional cytoplasmic domain, whilst the filled-in boxes represent transmembrane and intracellular sequences common to the 180 and 140 kd forms. The smallest form (120 kd) is attached to the membrane via a phosphatidylinositol linkage (∽).

contrast, the smaller 120 kd form lacks a transmembrane domain (13,14), but is attached to the membrane via a phosphatidylinositol linkage, presumably associated with the unique C-terminal domain predicted for the smallest NCAM (14,15) (*Figure 1*). A second unusual and characteristic feature of NCAM is its high content of α-2−8-linked polysialic acid, a form of sialic acid not previously known to occur in higher vertebrates (16−18).

Polysialic acid (PSA) comprises as much as 30% by weight of NCAM from embryonic and early post-natal (post-hatch) brain. In the adult brain, the content of sialic acid is reduced to one-third this level. This large content of PSA causes embryonic brain NCAM to migrate as a broad heterodisperse band on SDS−polyacrylamide gels. Treatment with neuraminidase, which specifically removes sialic acid, dramatically reduces this heterogeneity, producing prominent bands of apparent molecular weights of 140 and 180 kd for embryonic and 120, 140 and 180 kd for adult brain NCAM (*Figure 1*). As can be seen in *Figure 1*, in addition to developmental changes in the level of PSA associated with NCAM, the pattern of expression of the three core glycoprotein forms also shows developmental modulation. During early stages of chick development (stages 5−25) only the intermediate, 140 kd, form of NCAM is present. After stage 25, the larger 180 kd core glycoprotein can be detected either by examining neuraminidase-treated samples on SDS–gel/immunoblots or using antibodies specific for the additional cytoplasmic domain that it contains (11,19). Interestingly, the expression of this larger form appears to accompany neuronal differentiation, and 180-kd-specific NCAM epitopes are abundant in differentiating neurons, but absent from cells in the NCAM-positive proliferative zones (11,20). Additionally, while other non-neural embryonic tissues express NCAM, they always contain the smaller 120 and 140 kd core polypeptides. At much later stages of neural development, the 120 kd form becomes prominent. Overlain on these general trends are variations in PSA content and polypeptide forms among different regions of the nervous system (21) and between cell soma and neuritic process (22).

2.1.2 Structure–function relationships

NCAM is thought to produce cell–cell adhesion via homophilic binding: NCAM molecules on apposing cell surfaces bind directly to one another to form adhesive bonds. Rutishauser et al. (23) have carried out several sorts of experiments to demonstrate this. First, affinity-purified NCAM, either in a soluble state or incorporated into reconstituted lipid vesicles, binds directly to NCAM-rich cells but not to NCAM-poor cells. If target cells are pre-treated with monovalent anti-NCAM Fabs and then washed free of unbound antibody, the binding of soluble NCAM or NCAM vesicles is inhibited. Secondly, if lipid vesicles containing purified NCAM are incubated alone with shaking, they form large aggregates. Addition of anti-NCAM Fabs to the incubation prevents this aggregation. Finally, soluble affinity-purified ^{125}I-labeled NCAM binds to NCAM–Sepharose, binding can be inhibited by addition of unlabeled NCAM. Cunningham et al. (24) localized this homophilic binding activity to a 65 kd proteolytic fragment, Fr1, containing the extreme N-terminal (extracellular) portion of NCAM. While no binding of ^{125}I-labeled Fr1 to retinal cells was detected under conditions in which intact NCAM bound, plastic beads carrying attached F1 would readily adhere to retinal cell surfaces. Unlabeled Fr1 also effectively competed with [^{125}I]NCAM for binding to retinal cells.

Recently, Cole and Glaser (25) have reported that a smaller, 25 kd, N-terminal subtilisin fragment of NCAM demonstrates homophilic binding. This fragment inhibits the binding of retinal cells to surfaces coated with NCAM and, when bound to surfaces, it can substitute for intact NCAM in mediating retinal cell binding. This group has also made the interesting observations that ^{3}H-labeled heparin binds to NCAM, and that heparin sulfate but not chondroitin sulfate inhibits the binding of retinal cells to NCAM-coated surfaces (26). The heparin-binding domain is also associated with the 25 kd N-terminal fragment of NCAM (25,27). Since pre-treatment of retinal cells with anti-NCAM Fabs is sufficient to inhibit their subsequent binding to NCAM on a second surface (23,25), heterophilic NCAM – heparan sulfate proteoglycan binding does not appear to be sufficient to mediate cell adhesion. This series of observations led this group to suggest that cell-surface-associated heparan sulfate proteoglycans may be important activators of NCAM-mediated homophilic binding (28). An alternative view of these results would be that the inhibition of homophilic NCAM – NCAM binding by heparan sulfate is analogous to the effect of anti-NCAM antibody binding. Thus NCAM – heparan sulfate interactions *in vivo* could inhibit rather than promote NCAM-mediated adhesions. Clearly, the prevalence of heparan sulfate proteoglycans on cell surfaces and in ECM make this an important area for further study.

The majority (90%) of sialic acid associated with NCAM is attached to a 42 kd CNBr peptide positioned between the N-terminal homophilic binding domain and the C-terminal cell-associated region (29). Sialic acid is not required for homophilic binding, and neither the 25 nor the 65 kd fragments discussed above contains sialic acid. Indeed, the converse is true (30,31). Lipid vesicles prepared with PSA-rich embryonic brain NCAM aggregate 3- to 4-fold more slowly than do vesicles prepared with adult brain NCAM. Removal of PSA by neuraminidase increases the rate of aggregation for embryonic brain NCAM – lipid vesicles by 3.6-fold but has little effect on adult brain NCAM-containing vesicles (30). The ability of monoclonal antibody 15G8 to recognize sialic acid associated with embryonic but not adult brain NCAM (29) suggests that qualitative as well as quantitative differences in the PSA content of embryonic versus adult brain NCAM may also be important in regulating the adhesion of NCAM-rich membranes. By modulating the effectiveness of NCAM-mediated adhesion, changes in PSA levels could have important functional effects. For example, reduced levels of PSA associated with NCAM late in neural development may act to stabilize neural contacts (19).

Variation in the intracellular domain of NCAM could obviously alter cellular responses to NCAM adhesion. The intracellular domain of NCAM is relatively large (362 amino acids for the 180 kd form) and could interact with second messenger systems or cytoskeletal components. In support of the latter idea, Schachner's group has used photobleaching techniques to study the lateral mobility of NCAM. They observed similar mobilities for the 140 kd form of NCAM as other membrane proteins but, in contrast, found the 180 kd form had a greatly reduced mobility. Corroborating this result, they also found that fodrin, a key submembranous cytoskeleton component, binds to the 180 kd NCAM polypeptide but not to the 120 or 140 kd forms (32). Unlike the 140 and 180 kd forms, the smallest NCAM polypeptide has no cytoplasmic or transmembrane domain (13 – 15) (*Figure 1*). Its attachment to the cell surface via a phospholipase-C-sensitive phosphatidylinositol linkage suggests that the adhesion of cells bearing this type of NCAM could rapidly be altered by activation of phospholipase C enzymes. Released NCAM could then be incorporated

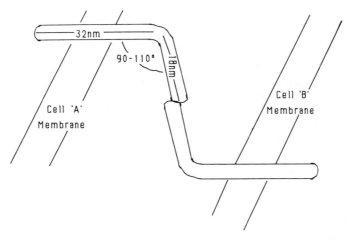

Figure 2. Schematic model of an 180 kd NCAM-dimer mediating intercellular adhesion based on electron microscopic visualization of purified chick NCAM. In addition to dimers, oligomers produced by the extracellular adhesion of three or four NCAM molecules may also occur. (After ref. 33.)

into ECM or could act as a competitive inhibitor of NCAM-mediated adhesion. Variations in either the sialylation or polypeptide form of NCAM may thus have important functional consequences on cell–cell adhesion.

Hall and Rutishauser (33) have used rotary shadowing techniques to visualize purified NCAM molecules in the electron microscope. Images could be oriented by the binding of specific monoclonal antibodies to previously identified epitopes. Individual molecules appeared as uniformly thick 4–5 nm rods containing a flexible hinge region in the extracellular domain. In addition, oligomers containing 2–6 such bent rods in a pinwheel-like, non-overlapping configuration were observed. Dimers represented the most frequent image. As indicated in *Figure 2*, the point of this homophilic association was located at the distal, extracellular end of the molecule. Enzymatic removal of PSA had no effect on the electron microscopic appearance of molecules, but increased the frequency of dimers and trimers, confirming the inhibitory effect of PSA on NCAM–NCAM association.

2.2 L1/NgCAM—a second calcium-independent neural adhesion molecule

Initially it was thought that glial cells lack NCAM and that other molecules would presumably be involved in neuron–glia cell adhesion (34). [In fact, NCAM is now known to occur on glial cells (35–37) and anti-NCAM Fabs have more recently been reported to inhibit adhesion of post-natal mouse cerebellar neurons and astrocytes (38).] To identify specific neuron–glial adhesion molecules, Grumet and co-workers in Edelman's laboratory developed heterotypic binding assays using suspensions of embryonic chick brain neurons and monolayers of non-neuronal, presumably glial, cells prepared from dissociated embryonic chick brain. To prevent neuron–neuron aggregation, Fab fragments of anti-NCAM antibodies were included in the assay while Fab fragments prepared from polyspecific anti-brain membrane antiserum were used to inhibit neuron–glial adhesion. In parallel with their previous isolation of NCAM, this assay was used to partially purify a fraction from chick brain which neutralized the effects of the anti-brain membrane antibodies on neuron–glial cell adhesion. This partially purified material was then used to generate monoclonal antibodies, two of which, 10F6 and 16F5, inhibited neuron–glial cell adhesion (39,40). When used to affinity-purify their respective antigen, both monoclonal antibodies gave preparations with a major component of 135 kd and minor bands of 200 and 80 kd. These antigens were given the descriptive label NgCAM, for neuron–glial cell adhesion molecule. While neither monoclonal antibody recognized purified NCAM (40), NgCAM prepared using these antibodies also contained NCAM (41), suggesting a possible interaction between these two adhesion molecules *in vivo*. The relationship between the three NgCAM components was tested by preparing antisera to each component and comparing their reactivities (41). Polyclonal antibodies prepared against the 200 kd component recognized all three polypeptides while anti-135-kd antibodies recognized the 135 and 200 kd, but not the 80 kd component. Similarly, anti-80-kd antibodies recognized only the 80 and 200 kd polypeptides, but not the 135 kd component. Thus the 80 and 135 kd components are each immunologically related to the larger 200 kd component, which may represent their precursor, but not to each other. Polyclonal antibodies to NCAM did not recognize any of the three purified NgCAM components, but as considered further in Section 2.5, monoclonal antibodies recognizing an N-linked carbohydrate epitope of NCAM also bind to NgCAM (41,42).

An adhesion molecule, L1, purified from post-natal mouse brain (43) appears to be analogous to chicken NgCAM. L1 was identified as a cell surface antigen recognized by a monoclonal antibody prepared using a glycoprotein-enriched fraction of cerebellar membrane proteins. The affinity-purified antigen from mouse cerebellum contains two prominent bands of 140 and 200 kd. Fab fragments prepared from monospecific polyclonal antibodies to L1 partially inhibit the calcium-independent aggregation of dissociated mouse cerebellar cells. As for NgCAM, the polypeptide portions of L1 and NCAM are immunologically distinct (44). Both L1 and NgCAM are found only on post-mitotic neurons. Unlike NCAM, neither are found in proliferative zones nor in association with CNS glial (41,43,44). L1/NgCAM is present on Schwann cells in the peripheral nervous system (45,46). Although no direct immunological or biochemical comparisons of L1 and NgCAM have been reported, both molecules are immunologically identical to the nerve growth factor-inducible large external (NILE) glycoprotein (47,48) originally characterized in the PC12 rat pheochromocytoma cell line (49).

Both the Edelman and Schachner groups have compared the ability of anti-L1/NgCAM and anti-NCAM antibodies to inhibit calcium-independent homotypic neuron – neuron and heterotypic neuron – glial cell adhesion. Using embryonic chick cells, Grumet et al. (41) observed no inhibition of neuron binding to glial monolayer cultures by anti-NCAM Fabs but anti-NgCAM Fabs reduced binding by about 50%. When these antibodies were tested in neuron aggregation assays, polyclonal antibodies to both NCAM and NgCAM inhibited aggregation of dissociated brain neurons. A mixture of the two antibodies was somewhat more effective than either alone. Interestingly, an anti-NgCAM monoclonal antibody (10F6) which inhibited neuron – glial cell adhesion had no effect on neuron – neuron adhesion, suggesting the mechanism of NgCAM-mediated adhesion differs between neuron – neuron and neuron – glial cell pairs. More recently, Grumet and Edelman (50) have used NgCAM-coated microscopic beads (Covaspheres) and NgCAM liposomes to demonstrate that NgCAM binds homophilically to NgCAM on neurons but heterophilically to an unidentified ligand on astroglial cells. Schachner's group, using mouse cerebellar neurons and astrocytes, found antibodies to NCAM disrupted both neuron – neuron and neuron – astrocyte adhesion. In contrast to findings with anti-NgCAM, polyclonal anti-L1 Fabs inhibited only neuron – neuron adhesion, no effects were seen on neuron – astrocyte adhesion (38). The discrepancy in these two groups' observations remains unresolved. Both observe significant inhibition of neuron – neuron adhesion by L1/NgCAM. The results of Grumet and Edelman (50) suggest that the differences in the effect of anti-L1/NgCAM antibodies on neuron – glial adhesion may reflect different specificities of the polyclonal antibodies used. Alternatively, there may be additional neuron – glia adhesion mechanisms in the rodent system which mask L1-mediated adhesion.

2.3 Myelin-associated glycoproteins

Myelin, a compact wrapping of glial cell membranes around axons, is a structural feature key to the functioning of both the central and peripheral vertebrate nervous systems. Myelination requires both heterotypic adhesion of the glial cell to the underlying axonal surface as well as homotypic tight adhesions in compact myelin between layers of oligodendrocyte membranes in the CNS or Schwann cell membranes in the PNS. The 25 – 30 kd oligodendrocyte proteolipid protein, named for its unique highly lipid-like character, and the Schwann cell 28 – 30 kd P_0 protein are thought to mediate this later homotypic adhesion (51,52,53). The proteolipid protein and P_0 each constitute about 50% of the total protein in CNS and PNS myelin, respectively.

A third myelin protein called MAG (myelin-associated glycoprotein), is produced by both oligodendrocytes and Schwann cells. It comprises approximately 1% of both CNS and PNS myelin and is thought to mediate heterotypic neuron–glia adhesion (54). Indirect support for this idea includes the localization of MAG to the periaxonal Schwann cell membrane; its absence from cell membranes within compact myelin (55); and an inverse correlation between the immunohistochemical localization of MAG and abnormally widened periaxonal spaces between myelinating Schwann cell and axon surfaces in the dysmyelinating mouse mutant, *Quaking* (54). Regions of normal axon–Schwann cell contact with the characteristic 12–14 nm spacing all contain MAG. Areas showing a widened spacing lack immunoreactive MAG. Schachner's group has now confirmed (56) this long-standing bias that MAG mediates cell–cell adhesion by showing that a monoclonal antibody directed against MAG can decrease neuron–oligodendrocyte and oligodendrocyte–oligodendrocyte adhesion in an *in-vitro* binding assay. Additionally, liposomes containing immunoaffinity-purified MAG bound specifically to neurites growing in culture. Since neurons do not contain detectable amounts of MAG, MAG-mediated heterotypic neuron–oligodendrocyte adhesion must involve heterophilic adhesion.

2.4 NCAM, L1/NgCAM, MAG, and P_0 have homologous extracellular domains

The complete nucleotide sequences for all three forms of chicken NCAM (9,14) and for the 120 kd and 140 kd forms of mouse NCAM (13,57) have recently been reported. The predicted amino acid sequences show 86% homology between the two species. The amino terminal (extracellular) portion contains five contiguous domains with sequence homology with each other and lower but significant homology with members of the immunoglobulin superfamily. Each domain consists of about 100 amino acids and contains a pair of cysteines spaced 50–56 amino acids apart. As NCAM contains no free cysteine nor interchain disulfides, these cysteines probably form intradomain disulfide bonds producing a series of five loops (*Figure 3*).

Interestingly, the predicted amino acid sequence of rodent L1 (58) and MAG (59–61) also contains a similar series of repeating domains, each having a pair of potentially disulfide-bonded cysteines spaced approximately 50–60 amino acids apart. The extracellular portion of L1 contains six such loops and MAG, five (*Figure 3*). As for NCAM these repeating domains show sequence similarity with each other and with members of the C2 subgroup (constant region like set 2) of the immunoglobulin superfamily. Other proteins in this group include the lymphocyte adhesion molecules ICAM (62), LFA3 and CD2 as well as the immunoglobulin F_c-receptor and the platelet-derived growth factor receptor (63). The myelin protein P_0 also contains an immunoglobulin-

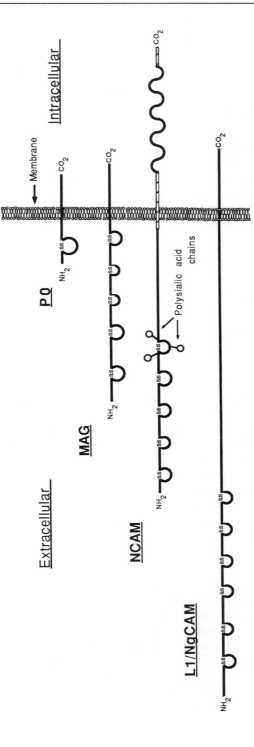

Figure 3. Schematic representation of the primary structure of P_0, MAG, NCAM, and L1/NgCAM. For simplicity, the extracellular immunoglobulin-like domains are shown as disulfide-linked loops. These are in fact thought to form a much more complex paired β-pleated sheet fold. The remainder of the proteins are shown as proportionately long straight or wavy lines although little is actually known about the organization of these regions. The largest, 180 kd form of NCAM is shown. The wavy intracellular line represents that portion of NCAM which is unique to the 180 kd form while the hatched line represents domains common to 140 kd and 180 kd NCAM. Note that the intercellular distance across which these four proteins might mediate adhesion could be significantly different.

```
NCAM 225  A N L G Q S V T L V C D A - D G F P E P - - - T M S W T K D G E P I E N E E D E -8aa-
L1   253  A L Q G Q S L I L E C I A - E G F P T P - - - T I K W L H P S D P M P T D R V I Y -3aa-
MAG  251  A I E G S H V S L L C G A - D S N P P P - - - L L T W M R D G M V L R - E A V A E
P0    40  G A V G S Q V T L H C S F W S S E W V S D D I S F T W R Y Q P E G G R D A I S I F -28aa-

          S S E V T I R N V D K N D E A E Y V C I A E N K A G E Q D A S I H L K V F A K P  308
          N K T L Q L L N V G E E D D G E Y T C L A E N S L G S A R H A Y Y V T V E A A P  331
          S L Y L D L E E V T P A E D G I Y A C L A E N A Y G Q D N R T V E L S V M Y A P  325
          D G S I V T H N L D Y S D N G T F T C D V K N P P D I V G K T S Q V T L Y V F E  147
```

Figure 4. Sequence homology between NCAM, L1, MAG, and P_0. Single letter amino acid sequences from the third immunoglobulin-like domain of mouse brain NCAM (amino acids 225–308; ref. 13), mouse brain L1 (amino acids 253–331; ref. 58) rat brain MAG (amino acids 251–325; ref. 59) and the single immunoglobulin domain of rat P_0 (amino acids 40–147; ref. 53) are aligned to show maximal identity. Amino acids which are identical in two or more of the sequences are boxed.

like domain, but of the V-subgroup type (63). All of the immunoglobulin superfamily are thought to be involved in intermolecular recognition and adhesion. In several cases this recognition involves heterophilic binding between members of the superfamily (63).

The sequence identity between domains for NCAM, L1, and MAG is greater when compared between each other rather than with other members of the immunoglobulin superfamily. *Figure 4* illustrates the aligned amino acid sequences for the third homology unit of these three adhesion molecules as well as the single domain of P_0. Pair-wise comparisons between the sequences shows a 32–34% sequence identity between NCAM, L1 and MAG and a 15–17% identity between each of these and P_0. The evolutionary relationships between members of this superfamily will undoubtedly be intriguing. The recent identification of neural cell surface proteins in insects with sequence similarity to NCAM, MAG, and L1 (64,65) strengthens the commonly accepted view that the earliest roles of the immunoglobulin superfamily proteins were in cell–cell recognition rather than in vertebrate immunity.

The individual homology domains of the immunoglobulin superfamily are thought to form a characteristic fold consisting of two β-pleated sheets forming a hydrophobic region between them and exposed hydrophilic surfaces. In immunoglobulin molecules, interdomain interactions occur between these hydrophilic surfaces. Although the precise role of homologous domains in other superfamily proteins is not as clear, these characteristic structures will likely participate directly in adhesion. For NCAM, the heparin- and homophilic-binding domains have been localized to the second and third immunoglobulin domains, respectively (9,66). Recently, Akeson's group (67) has reported an NCAM polypeptide variation expressed by a subset of neurons resulting from the addition of 10 amino acids in the middle of the fourth immunoglobulin-like domain of NCAM. They note that this insertion aligns with the position of the hypervariable regions of immunoglobulin protein variable domains. Were such variability to result in altered binding activities it would certainly add a new dimension to the biology of NCAM.

2.5 Calcium-dependent neural adhesion molecules

Early studies of retinal cell adhesion had identified two functionally distinct adhesion systems, one which was independent of the presence of calcium in the adhesion assay, and a second which required calcium (68–71). The latter system could be selectively preserved on cell surfaces by moderate trypsinization in the presence of calcium. NCAM is a major contributor to calcium-independent neural cell adhesion (L1/Ng CAM- and MAG-mediated adhesion is also calcium-independent). Using techniques analogous to those employed to identify NCAM, Lilien's group have identified a molecule involved in calcium-dependent neural cell adhesion (72). Polyspecific antisera were prepared against retinal cells dissociated using the trypsin/calcium technique (TC cells) which show only calcium-dependent adhesion. While all such antisera bound to TC cells, monovalent Fabs from only a subset of these inhibited adhesion of TC cells. One such blocking antiserum was used to immunoprecipitate radiolabeled cell surface proteins from TC cells and from adhesion-defective retinal cells prepared by trypsinization in the absence of calcium (i.e. trypsin/EDTA or TE cells). Comparison of two-dimensional gel electrophoretic profiles revealed two polypeptides, 70 and 130 kd, which were immunoprecipitated from TC but not TE cells. As for NCAM, a factor released by retinal cells in culture was then purified on the basis of its ability to neutralize the adhesion-blocking activity of the anti-TC cell antibodies. The resulting polypeptide was shown to be immunologically related to the TC-cell-associated 70 and 130 kd proteins as it specifically competed with them for antibody binding in immunoprecipitation experiments. Since the 70-kd cell-associated protein appeared only after trypsinization of cells (73), the larger 130 kd protein was considered to be the native form of the calcium-dependent adhesion molecule, which has been referred to as N-cal-CAM by this group. Hatta and Takeichi (74) have used monoclonal antibody techniques to identify what appears to be the same calcium-dependent adhesion molecule described by Lilien and co-workers. Several monoclonal antibodies were obtained which reacted with TC chick neural retinal cells but not those trypsinized in the absence of calcium. All of these antibodies recognized a 127 kd retinal cell surface protein. Additionally, one monoclonal antibody, NCD-2, inhibited the calcium-dependent adhesion of TC cells but had no effect on the calcium-independent adhesion of cells prepared with low trypsin and EDTA (LTE cells). This 127 kd calcium-dependent adhesion molecule has been named N-cadherin. As with N-CAM, N-cadherin is present on a variety of embryonic tissues early in development but becomes restricted to the nervous system in the adult (75).

N-cadherin is now known to be one of a family of glycoproteins including E-cadherin [also known as LCAM (liver cell adhesion molecule) and uvomorulin] and P-cadherin, which all mediate calcium-dependent cell–cell adhesion (76). These molecules, which were named after the tissue from which each was initially isolated (neural, epithelial, and placental, respectively), are expressed in unique tissue distribution patterns and appear to be the principle adhesive components of cellular adherens junctions. Takeichi's group has recently determined the predicted amino acid sequence for these three cadherins (76,77). They each appear to have a single transmembrane domain with about three-quarters of their sequence external to the membrane. More importantly, the three cadherins show 50–60% sequence identity throughout both their extracellular and intracellular domains. No significant sequence similarity was found with other known proteins. Thus NCAM-L1/NgCAM – MAG – P_0 and the cadherins represent two distinct families of adhesion molecules. A third family consists of a group of receptors for ECM molecules called cytoadhesions (described in Section 2.8).

2.6 J1/cytotactin and the L2/HNK-1 family of cell adhesion molecules

HNK-1, a monoclonal antibody produced against a membrane antigen of a lymphocyte cell line, reacts with a specialized subset of T cell lymphocytes including natural killer and antibody-dependent killer cells (78). Additionally, the antibody was noted to react with antigens present in the central and peripheral nervous systems (79), including the 100 kd myelin-associated glycoprotein (MAG) (80). HNK-1 recognizes an unusual sulfated glucuronic acid determinant (81,82) present on several glycoproteins in both the developing and adult nervous system and transiently on glycolipids during development (82). The HNK-1 determinant appears to be the major antigen recognized by characteristic MAG-reactive IgM antibodies associated with a group of human peripheral demyelinating neuropathies (81,83,84).

Several other, independently isolated, carbohydrate-dependent monoclonal antibodies (L2, NC-1, and 4F4), obtained following immunizations with neural tissue, were noted to stain similarly to HNK-1 and have all been shown to have identical or nearly identical immunological reactivities (42,82,85). Schachner's group made the interesting observation that in addition to MAG, the two other principle glyco-proteins recognized by L2/HNK-1 were NCAM and L1/NgCAM (42). Interestingly, L2/HNK-1 was noted to only react with a subset of NCAM and L1/NgCAM molecules, and to stain only a subpopulation of NCAM- and L1/NgCAM-positive neurons as well as subpopulations of astrocytes and oligodendrocytes.

Since the three major glycoproteins recognized by L2/HNK-1 (NCAM, L1/NgCAM, and MAG) are all thought to be neural adhesion molecules,

Schachner's group reasoned that L2/HNK-1 antibodies may identify a family of adhesion molecules (42). In addition to MAG, NCAM, and L1/NgCAM, the other major component recognized in adult mouse brain by L2/HNK-1 is a glycoprotein of about 160 kd. To test the prediction that this molecule, termed J1 by Schachner's group, also mediates adhesive interactions, monospecific polyclonal antibodies were prepared against it. These antibodies stain the surface of astrocytes and oligodendrocytes but not neurons in monolayer cultures of dissociated postnatal mouse cerebellum. Monovalent anti-J1 Fabs block the heterotypic adhesion of purified cerebellar neurons and astrocytes but not homotypic neuron – neuron or astrocyte – astrocyte binding (86). Thus it appears that all of the major glycoproteins recognized by L2/HNK-1 in the adult nervous system may be involved in cell – cell adhesion. Monoclonal anti-L2 antibodies also partially inhibit neuron – neuron, neuron – astrocyte and astrocyte – astrocyte adhesion (38). Whether this inhibition reflects the direct involvement of the L2/HNK-1 carbohydrate determinant in adhesion or steric inhibition of a nearby but distinct binding domain remains to be tested.

The avian equivalent of J1 has been isolated using a very similar approach by Grumet and co-workers in Edelman's laboratory (87). An antiserum was prepared against a mixture of NC-1 (an HNK-1-like monoclonal antibody)-reactive glycoproteins and then absorbed with purified NCAM and NgCAM. Fab fragments of these antibodies inhibited neuron – glial adhesion, and were used to monitor the purification of HNK-1 antigens which could neutralize this inhibition. The resulting HNK-1-reactive antigen, isolated from embryonic chick brain, had a molecular weight similar to J1 and was also present on glia but not neurons. Additionally, Grumet et al. (87) noted that the factor was present in ECM material in smooth muscle, lung and kidney. Thus, in view of its potential role in mediating both cell – cell and cell – substrata adhesion the name cytotactin was chosen. Saturating amounts of anti-cytotactin antibodies inhibited neuron – glial adhesion as effectively as did anti-NgCAM antibodies (39 versus 40%). A combination of both antibodies inhibited adhesion to a greater extent (63%) suggesting cytotactin and NgCAM represent separate molecular mechanisms for neuron – glial adhesion in this system.

J1/cytotactin appears to be a secreted glycoprotein capable of interacting with ECM molecules. Most of the antigen can be solubilized from brain in the absence of detergent and in primary cultures the majority of antigen produced is found in the culture supernatant. In both neural and non-neural tissues, J1/cytotactin is present as a high molecular weight disulfide-bonded complex. Following reduction, J1/cytotactin from embryonic brain migrates as components of 220, 200, and 190 kd. In adult brain, anti-J1 antibodies predominantly recognize a 160 kd band. Although J1/cytotactin has a similar molecular weight to fibronectin (220 kd), it is immunologically distinct from fibronectin, laminin, collagen type IV, and heparan sulfate proteoglycan (86–88). A previously described chick myotendinous antigen (88), thought to participate in the adhesion of muscle fibers to tendons, appears to be identical to J1/cytotactin. This antigen is secreted by primary myogenic cultures as a family of 220, 200/190, and 150–170 kd glycoproteins. Limited proteolytic digestion of these individual polypeptides yields similar cleavage patterns suggesting a close structural relationship between the multiple antigen forms. Chiquet and Fambrough (88) noted that the myotendinous antigen is non-covalently complexed to a chondroitinase-sensitive, very high molecular weight proteoglycan-like material suggesting the antigen may form a linkage between muscle fibers and tendon-associated ECM by binding to proteoglycans on the muscle fiber surface.

Hoffman and Edelman have recently implicated a similar proteoglycan in the binding of neurons to cytotactin (89). They have identified and characterized a chondroitin sulfate proteoglycan complex synthesized by organ cultured brain but not cultured glial cells, which co-purifies with cytotactin during immunoisolation on HNK-1 antibody–agarose columns. The complex contains hyaluronic acid and a 280 kd core protein that surprisingly also contains the HNK-1 epitope! Using Covaspheres coated either with this proteoglycan or with cytotactin, specific adhesive interactions between the two molecules could be demonstrated. Neither cytotactin beads nor proteoglycan beads formed homotypic aggregates, but when incubated together, they readily formed aggregates. This heterotypic binding was inhibited by either soluble cytotactin or the purified proteoglycan as well as by anti-cytotactin antibodies. Laminin, which is known to form complexes with proteoglycans in other systems, also inhibited heterotypic bead aggregation. Addition of fibronectin to the assay had no effect. In a series of similar binding experiments which also included laminin- and fibronectin-coated beads, specific cytotactin–proteoglycan, laminin–proteoglycan, and cytotactin–fibronectin binding was observed. When the binding of cytotactin-coated beads to neurons was studied, anti-cytotactin antibodies, purified chondroitin sulfate proteoglycan and laminin all inhibited cytotactin–neuron adhesion. In contrast, a heparan sulfate proteoglycan from Engelbrecht–Holm–Swarm sarcoma cells was not inhibitory. Thus, J1/cytotactin has multiple binding activities which could allow it to bind both to cell surfaces and ECM components. Based on these observations, Hoffman and Edelman (89) suggest that cytotactin, absorbed to the surface of glial cells, produces neuron–glial adhesion by specific interaction with a neuronal cell surface chondroitin sulfate proteoglycan. Interestingly, the proteoglycan core protein contains the HNK-1 epitope, extending the family of molecules involved in cell adhesion which carry this uncommon carbohydrate.

2.7 AMOG—adhesion molecule on glia

AMOG was discovered in the same manner as L1, during a continuing series of hybridoma experiments intended to identify and characterize cerebellar cell surface glycoproteins. The AMOG antigen is a 50 kd glycoprotein present on the surface of murine cerebellar astrocytes and oligodendrocytes, but not cerebellar neurons (90). Nothing has been reported about the distribution of AMOG outside the cerebellum. Monoclonal anti-AMOG antibodies have little effect on astrocyte – astrocyte adhesion, but inhibit the calcium-independent adhesion of cerebellar neurons and astrocytes by about 25%. In the latter adhesion pair only the astrocyte contains AMOG, thus, if directly involved in binding, AMOG-mediated neuron – glial adhesion would be heterophilic.

2.8 Extracellular matrix adhesion molecules

The ECM components collagen, fibronectin, and laminin have long been known to promote the adhesion of dissociated cells to tissue culture vessels. Of these, considerable attention has been given to study of the role of fibronectin and laminin in neural cell migration and axon growth. This section briefly reviews the structure of these molecules and then considers recent advances in the identification of cell surface receptors for ECM components.

2.8.1 Fibronectin

Fibronectins exist as a family of ECM, cell surface, and soluble glycoproteins (typically composed of a disulfide-bonded 440 kd dimer) all derived from a single gene by alternative RNA splicing (91 – 94). Plasma fibronectin is synthesized principally by the liver and has important functions in hemostasis/thrombosis and phagocytosis. Cellular fibronectins contain additional peptide sequence in the internal regions of the molecule (91) and are synthesized by a number of cell types including fibroblasts, myogenic cells and glial cells. *In vitro*, fibronectin can influence the adhesion, differentiation and migration of a number of cell types (see ref. 95 for a review).

Fibronectin is a highly extended molecule composed of a series of repeating amino acid homology units (reviewed in ref. 96). Together these form multiple, functional domains mediating fibronectin's binding to other ECM molecules including collagen, fibrin, and heparin, and to cell surface receptors. Each of these binding domains have been mapped along the molecule by identifying proteolytic fragments which retain individual activities. Guided by the primary amino acid sequence of the cell binding domain, Pierschbacher and Ruoslahti's group have produced synthetic peptides which mimic the cell binding properties of fibronectin (97 – 100). Peptides containing the sequence Arg-Gly-Asp (RGD) promote fibroblast cell attachment when bound to surfaces and inhibit the attachment of fibroblasts to fibronectin-coated dishes in solution. As discussed further in the next section, the RGD sequence is also important for the binding of a number of molecules to cell surface receptors.

For some cell types, recent studies indicate the presence of alternative or additional cell binding activity in the C-terminal heparin-binding domain of fibronectin. For example, while proteolytic fragments containing the cell-binding domain promote cell–substratum attachment and neurite outgrowth from PNS neurons, neurite outgrowth from CNS neurons is not stimulated by this region. In contrast, fragments derived from the C-terminal heparin-binding region promote cell binding and process growth for both classes of neurons (101). Similarly, Humphries et al. (102) found that the herapin-binding region, but not the internal fibroblast cell-binding domain, mediates the adhesion and spreading of another neural ectoderm derivative, melanoma cells. These workers identified a tetrapeptide sequence Arg–Glu–Asp–Val (REDV) present in this C-terminal region which appears to be involved in melanoma cell binding to fibronectin since synthetic peptides containing the REDV sequence inhibit melanoma but not fibroblast adhesion and spreading on fibronectin. Interestingly this sequence is carried in a region of fibronectin involved in differential exon splicing, suggesting that cell-specific expression of fibronectin receptors having different binding requirements together with the regulated synthesis of fibronectins containing alternate cell binding domains could produce multiple adhesive specificities.

2.8.2 Laminin

Like fibronectin, laminin can promote the adhesion of a number of cell types (principally epithelial) to culture dishes. In addition, interaction with laminin promotes the survival and differentiation of a variety of cells (for reviews see refs 103,104). Laminin, a principle component of all basal lamina, was first isolated in a soluble form from the mouse EHS sarcoma (105). In its native form laminin consists of three disulfide-bonded subunits (A, 440 kd; B1, 225 kd; B2, 205 kd) which form an asymmetric cross-like structure (106). Molecular cloning and sequence analysis indicates these three subunits arise from related genes since they share extensive regions containing 20–40% amino acid sequence identity (107). Laminin binds each of the other major components of basal lamina: collagen type IV (108); entactin/nidogen (109); and heparan sulfate proteoglycan (110). It is thought to be central to the structural organization of basal lamina. In addition, laminin contains at least two cell-binding domains (111–115).

In vitro, laminin is a potent inducer of neurite outgrowth for both central and peripheral neurons (116–119) and enhances the effect of soluble neurotrophic factors (117,120) on neuron survival. Factors produced by a number of non-neuronal cells when bound to culture dishes, show similar effects on cultured neurons. Purification of these factors has shown laminin, complexed with a heparan sulfate proteoglycan, to be the principle active component (121,122). Both the survival- and neurite outgrowth-promoting effects of laminin have been mapped to the heparin-binding domain on the long arm (A subunit) of laminin (117), an area not previously thought to be involved in cell binding (but see ref. 111). Edgar's suggestion

(117) that neurons interact via cell surface heparan sulfate proteoglycans with the heparin-binding domain of laminin would provide an interesting parallel with Hoffman and Edelman's (89) proposal that neuron binding to J1/cytotactin also occurs via a cell surface proteoglycan.

Recently, a novel laminin-like 190 kd basal lamina protein, S-laminin, has been identified (107). S-laminin is selectively concentrated in the synaptic cleft of nerve–muscle synapses. Molecular cloning and sequence analysis of S-laminin revealed an unsuspected 50% amino acid sequence identity with the B1 subunit of laminin. Neurons selectively bind to purified or recombinant S-laminin to an extent comparable with laminin. In contrast, neurites growing *in vitro* appear to stop upon contact with a patch of S-laminin (D.Hunter, personal communication). This result is particularly interesting in view of the striking tendency of regenerating motor axons to selectively recognize and produce differentiated nerve terminals at S-laminin-rich former synaptic sites (123,124). Thus the A, B1, and B2 laminin subunits and S-laminin may represent another family of related adhesion–recognition molecules which could help regulate complex patterns of axon growth.

2.8.3 *Cell surface receptors for laminin and fibronectin*

A 68 kd cell surface protein has been isolated from several murine and human cell lines by affinity chromatography on laminin–Sepharose (125–128; see also ref. 104). The purified protein shows a high affinity for laminin ($K_d = 2 \times 10^{-9}$ M; ref. 127) and presumably contains a membrane-binding domain since the protein can be reconstituted into liposomes which specifically bind to laminin (125). Antibodies prepared against the purified protein inhibit laminin binding to intact MCF-7 mammary carcinoma cells confirming that the 68 kd polypeptide is a cell surface laminin receptor. Unfortunately, with respect to the current chapter, no studies have been carried out to identify this receptor in neural cells.

Cell surface receptors for fibronectin have been identified directly by affinity chromatography on 'fibronectin–Sepharose' (129) and indirectly by the study of antigens recognized by monoclonal antibodies selected by their ability to disrupt cell–substratum adhesion (130–132). In the first approach, Pierschbacher and Ruoslahti's group (129) made an affinity matrix using the cell-binding fragment of fibronectin which they had previously identified (99). A fibronectin receptor complex present in detergent extracts of human osteosarcoma cells was specifically eluted from the column using synthetic peptides containing the key RGD triplet recognition sequence of the fibronectin cell-binding domain (98,100). Reconstituted liposomes containing the purified receptor bound to fibronectin, but not to laminin-coated surfaces, and binding could be inhibited by addition of RGD-containing peptides. Under reducing conditions, the receptor migrated on SDS–gels as a 140 kd band. In the

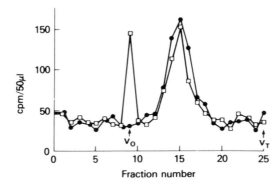

Figure 5. Equilibrium gel filtration assay of laminin – CSAT antigen binding. Laminin and ^{35}S-labeled CSAT antigen were applied to an Ultrogel AcA22 column pre-equilibrated with 400 µg ml^{-1} laminin and 1 mg ml^{-1} of the fibronectin cell-binding tetrapeptide, Arg-Gly-Asp-Ser (RGDS) (□) or with laminin and an irrelevant peptide (●) and the elution profile determined. For columns pre-equilibrated with buffer alone, CSAT antigen elutes as a single peak centered at fraction 15. Interaction of the CSAT antigen with laminin produces a new elution peak at the void volume, V_o. The RGDS tetrapeptide specifically inhibits laminin – CSAT antigen interaction. V_T indicates the total column volume. (Reproduced with permission from ref. 134.)

second, less direct approach, two independently isolated monoclonal antibodies, JG22 (131) and CSAT (132), which disrupt the adhesion of primary chick myoblasts to gelatin-coated dishes, were both found to recognize a similar fibronectin receptor (133,134). Binding of the CSAT/JG22 antigen to fibronectin was demonstrated using Hummel – Dryer (135) equilibrium-binding gel filtration assays which are useful for study of relatively low affinity interactions. Additionally, the antigen also binds to laminin in these assays and, like fibronectin, binding is inhibited by peptides containing the RGD triplet (*Figure 5*; 134). Confirming this result, both the CSAT antibody and RGD-containing peptides were shown to inhibit the adhesion of chick tendon fibroblasts to either laminin or fibronectin surfaces (134). In contrast, the adhesion of cardiac fibroblasts to laminin was not affected, suggesting the presence of additional laminin receptors (perhaps the 68 kd laminin receptor) in these cells. The CSAT/JG22 antigen additionally binds talin, an actin filament-binding cytoskeletal component (136), suggesting the receptor is intimately involved in providing a transmembrane linkage between ECM components and the cytoskeleton (137). In light of this linkage property, the name integrin has been proposed for the CSAT/JG22 antigen complex (138). As will be discussed further in Sections 3.2 and 3.4.1, integrin may be involved in neuron – ECM-binding since CSAT and JG22 antibodies block neuron binding and neurite outgrowth on ECM-coated surfaces (139,140).

The RGD tripeptide sequence recognized by the human fibronectin receptor and by integrin, the chick fibronectin/laminin receptor, appears to be a commonly used recognition sequence for a family of cell surface receptors (141,142). In addition to fibronectin, the RGD sequence is present in vitronectin, a serum-spreading factor (143), type I collagen (144), fibrinogen (145) and discoidin I, an adhesion protein of *Dictyostelium* (146). Binding of specific cell types to each of these adhesion molecules has been shown to involve the RGD sequence (147–150). By analogy with their purification of the fibronectin receptor (129), Pierschbacher and Ruoslahti's group have used RGD-containing peptides to elute distinct cell surface receptors from appropriate antigen columns for vitronectin (151), type I collagen (147), and a platelet receptor (gp IIb/IIIa) which binds fibronectin, fibrinogen, vitronectin, and the von Willebrand factor (152). While the RGD sequence is central to receptor binding to these proteins, the sequences flanking RGD differ between these proteins and, with the exception of the platelet receptor, each receptor appears to recognize the RGD sequence in only one protein. Thus, while using a common binding site, this group of receptors discriminates between the polypeptide context in which the RGD sequence is presented.

Comparison of available sequence data, polypeptide composition and immunological cross-reactivity for integrin, the vitronectin receptor and the platelet gpIIb/IIIa receptor, suggests these receptors are members of a larger family of cell surface adhesion/recognition complexes (141,142) which have been termed cytoadhesins (153). [Other authors have used integrin as a generic term for this superfamily but the term cytoadhesins prevents confusion between the superfamily label and the specific integrin receptor(s).] This group which also includes the leukocyte family of adhesion proteins (LFA-1, Mac-1, p150,95; ref. 154), and the human VLA (very late antigen) group of cell surface proteins (155) all consist of non-covalently bound dimers containing 130–210 kd α subunits and 95–130 kd β subunits. Integrin, which occurs in chick fibroblasts and myoblasts as three polypeptides, probably represents a pair of dimers sharing a common β subunit. [The CSAT and JG22 monoclonal antibodies both recognize this β subunit (156).] Recently, derived amino acid sequences for the β subunit of integrin (138), the LFA-1, Mac-1, p150,95 leukocyte glycoprotein family (154) and the platelet receptor gpIIb/IIIa (157) have been reported. All three are transmembrane proteins with large extracellular domains which contain four adjacent cysteine-rich domains (40 amino acids each). The sequences show significant homology (45–47% identity) including 56 identically positioned cysteine residues. Based on immunological comparisons (158,159), these three β polypeptides define three families of cytoadhesins which each share a common β subunit (141). One family includes integrin, the human fibronectin receptor, and the VLA group of antigens, a second contains the vitronectin and platelet gpIIb/IIIa receptor, and the third the leukocyte family of adhesion proteins. Within each family, members have distinct α subunits which also appear to share sequence homologies (160,161).

Representatives of the cytoadhesin family are widely distributed, occurring on nearly all vertebrate cell types including neurons and neural crest cells (162–164). The distribution of individual members can be widespread or highly restricted (155), and multiple cytoadhesins can be expressed by single cell types (151,155). In the nervous system, only the distribution of the integrin branch of the cytoadhesin family has been examined (162,163). Interestingly, in a study of the polypeptide composition of complexes immunoprecipitated from neural retina by antibodies to the β subunit of integrin, age-related changes in the pattern of α-like subunits were seen (140). As has been pointed out by other authors (141,142,165) the heterogeneity of the cytoadhesin family coupled with developmental and tissue-specific regulation of their expression, suggests that this group of glycoproteins may play an important role in determining the adhesive specificity of cells during morphogenetic processes. With respect to this idea, it is interesting to note the parallels between the cytoadhesins and a family of related cell surface antigens of *Drosophila* (165). The position-specific (PS) antigens, named for their differential spatial distribution on developing imaginal disk epithelia (166,167), share a common 110 kd β subunit but have distinct α subunits (168,169). Recently, comparison of N-terminal sequences for the α subunit of PS-1 with those for several cytoadhesin α subunits has shown significant homology (25–45%; ref. 161). In contrast to larval imaginal disks, within other larval and adult tissues, the PS antigens show only minor spatial restrictions, but instead exhibit tissue-specific patterns of expression (169). Hopefully, studies will soon be carried out to examine whether these cytoadhesin-like *Drosophila* cell surface molecules also mediate cell adhesion.

3. Involvement of cell adhesion molecules in neural development

In the following several sections, data will be reviewed which point to the involvement of specific adhesion molecules in neural cell interactions during development and regeneration. For many examples, the initial implication of particular adhesion molecules in specific cell interactions has been the result of guilt by association from immunohistochemical observations. Clearly, the careful histological demonstration of an adhesion molecule at sites of cell interaction is a minimal, and sometimes overlooked, requirement for models of function *in vivo*. For most of the examples presented, there have additionally been *in-vitro*, and in some cases *in-vivo*, functional studies to support the proposed role.

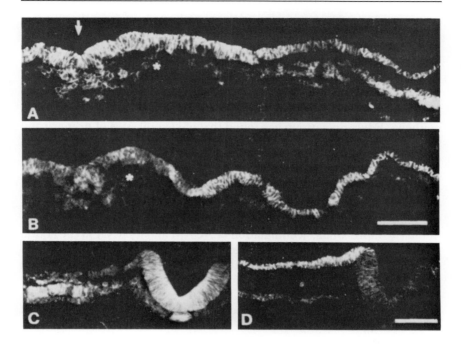

Figure 6. Reciprocal changes in the ectodermal distribution of NCAM and LCAM during neurulation. **(A)** and **(B)**, Transverse sections through the primitive streak (arrow) of stage 6 chick embryo stained with **(A)** anti-NCAM or **(B)** anti-LCAM antibodies. Both antibodies stain the entire ectoderm. **(C)** and **(D)**, Transverse-sections from five-somite (stage 8) chick embryo taken just caudal to the last formed somite stained with **(C)** anti-NCAM or **(D)** anti-LCAM antibodies. Anti-NCAM strongly stains the invaginating neural ectoderm and lateral mesenchymal cell layer, but now poorly stains ectodermal cells lateral to the edges of the neural groove. Anti-LCAM exhibits a reciprocal staining pattern from anti-NCAM in the neural and epidermal ectoderm. The bar is 100 μm. (Reproduced with permission from ref. 7.)

3.1 Modulation of cell – cell adhesion during primary neural induction

Development of the vertebrate nervous system begins with the formation of the neural tube. Following elaboration of the three primary layers during gastrulation, a longitudinal groove forms in the ectoderm. The edges of this groove, delimiting the neural ectoderm, elevate and eventually meet at the dorsal midline. Coalescence of these edges and separation from the lateral ectoderm gives rise to a neural tube overlain with a continuous ectodermal epithelia. Edelman's group has compared the distribution of two primary cell adhesion molecules, NCAM and LCAM, during this process and find that segregation of the neural ectoderm from the adjacent ectoderm may involve the regulated expression of these two CAMs. LCAM is a Ca^{2+}-dependent adhesion molecule initially isolated from chicken liver (170) and is equivalent to E-cadherin in rodents (171). Both NCAM and LCAM are present on derivatives of all three primary germ layers (7,172), but as development proceeds NCAM becomes increasingly restricted to neural tissues and LCAM to endodermal and ectodermal epithelia. Edelman's group has found that at several sites of primary and secondary induction, epithelia which initially express both CAMs lose one or the other as part of the inductive process (7). As shown in *Figure 6*, at the primitive streak stage in chick development, ectodermal cells express both LCAM and NCAM. As the neural tube forms from this epithelia, the neural ectodermal cells begin to express only NCAM, while the remaining ectodermal cell population begins to express only LCAM. Similarly, in *Xenopus*, accumulation of NCAM and its RNA in the neural plate ectoderm accompanies neural induction by the underlying mesoderm (173,174). Recently, Hatta and Takeichi (74) reported that during the period of neural tube closure the neuroectoderm begins the *de novo* expression of N-cadherin. Thus, spatial–temporal changes in the expression of the cell-adhesion molecules NCAM, LCAM, and N-cadherin may act to promote segregation of the neural ectoderm and formation of the neural tube.

3.2 Neural crest cell migration

The neural crest is a transitory structure which forms from the neural ectoderm along the dorsal–lateral ridges of the neural tube at the time of neural tube closure. Shortly after its formation, crest cells begin a migratory process along characteristic pathways to form the sensory and sympathetic ganglia of the peripheral nervous system (reviewed in refs 175,176). Other principal neural crest derivatives include the adrenal medulla and melanocytes. Largely, our understanding of the migratory pathway and ultimate fate of crest cells derives from study of quail-chick chimeras. Quail cell nuclei can be readily distinguished in these chimeras by their characteristic condensed heterochromatin, and following isotopic quail to chick transplants of neural tube segments, the migration and fate of neural crest cells can be studied. More recently, the migration of neural crest cells has been followed using monoclonal antibodies (NC-1 and HNK-1) which selectively stain the surface of migrating crest cells. The NC-1 antibody which was isolated following immunizations with ciliary ganglion cells (177), also stains most neuroectoderm derivatives later in development (178). As noted in Section 2.5, the antigen recognized by the NC-1 monoclonal antibody appears to be identical or closely related to the carbohydrate epitope carried by several neural cell adhesion molecules recognized by HNK-1 antibody (85). The neural crest cell molecule(s) containing the HNK-1/NC-1 antigen have not been characterized, but immunohistochemical staining with both NC-1 and HNK-1 have provided a more detailed description of initial neural crest cell migration pathways (179–181). At trunk levels, crest cells initially migrate ventrally into a cell-free space limited by the neural tube, somite, and epidermal ectoderm. In the electron microscope, this space contains a meshwork of thin filaments and interstitial bodies (182,183) both of which contain fibronectin and glycosaminoglycans (184). Further ventral crest cell migration is principally via intersomitic spaces and a pathway between the basal lamina enveloping the dermamyotome and the underlying sclerotome in the rostral portion of each somite. Several hours later a large number of NC-1/HNK-1 cells migrate laterally through the rostral portion of each sclerotome in conjunction with the invasion of ventral root axons (180,181). With the exception of the latter route, the initial migration of crest cells is through relatively cell-free spaces containing large amounts of fibronectin (162,163,185) and other ECM molecules, suggesting that ECM components are the principle substrates for crest cell migration.

In vitro, crest cells are readily identified as a population of cells which rapidly migrate away from the dorsal surface of neural tube explants. They readily attach to and migrate on fibronectin- but not collagen-coated surfaces (186). Although crest cells bind more slowly to laminin substrata (186), laminin also appears to be a suitable substrate for migration (187,188). Most neural crest cells fail to synthesize fibronectin (185) but do express the fibronectin/laminin receptor identified by CSAT/JG22 monoclonal antibodies (162,163). Treatment of cultures with antibodies to this receptor prevent neural crest migration on fibronectin and laminin substrates (162,187). To test whether interaction of crest cells with fibronectin and/or laminin is also necessary *in vivo* for their migration, antibodies against integrin (187,189) and synthetic peptides containing the cell binding RGD triplet of fibronectin (190) have been injected into chick embryos at sites of neural crest cell migration. Both treatments inhibited the initial migration of crest cells into the ventro–lateral cell-free spaces, resulting in a greatly reduced number of crest cells in ventral pathways and the anomalous accumulation of crest cell aggregates, first, within the lumen of the neural tube and secondly, dorsal to the neural tube. Injections of antibodies which bind to integrin but fail to block fibronectin binding *in vitro*, or of fibronectin peptides not containing the RGD sequence, had no effect on crest cell migration. The conclusion from these experiments was that fibronectin present in neural crest pathways is crucial to crest cell migration. However, since the RGD triplet is involved in cell binding to several ECM molecules and the anti-integrin antibodies used by Bonner-Fraser (187,189) react with the β subunit which may be common to other integrin family members, this conclusion must be regarded as tentative. The data do, however, indicate that adhesion of crest cells to ECM molecules is important for their early migration.

Little is known about subsequent adhesive or recognition mechanisms which guide crest cells along latter stages of their migration and aggregation into ganglia. The expanding somite is thought to produce a physical restraint to further migration for those cells which will give rise to the sensory dorsal root ganglia. In contrast, for those crest cells which penetrate ventrally past the somite to reach the aorta and gut, there is no readily apparent physical barrier or pattern of fibronectin/ECM which would limit their migration to sites of future autonomic ganglia (191). An alternative mechanism might be the induction of cell–cell adhesion molecules on crest cells by their contact with specific mesenchymal components. By comparing patterns of NC-1 and NCAM immunoreactivity, Duband et al. (192) find that induction of NCAM expression may be involved in the formation of sympathetic and parasympathetic ganglia. During active phases of migration, neural crest cells express relatively little NCAM, but as autonomic ganglia form (via aggregation of migrating cells or local proliferation), crest cells begin to exhibit bright staining with anti-NCAM antibodies. In contrast, migrating crest cells, but not those in newly formed sympathetic ganglia, are readily stained with anti-fibronectin receptor antibodies (162). The reciprocal expression of cell–substratum versus cell–cell adhesion molecules may thus be important for the formation of autonomic ganglia.

3.3 Cerebellar granule cell migration

Most neurons in the vertebrate CNS migrate from their site of birth (e.g. last mitosis) to their final location. In contrast to cell migration in most tissues, cortical neuron migration typically involves the intracellular translocation of the cell nucleus through a previously extended cellular process rather than a migration of the entire cell. For cerebellar granule cell neurons situated in the internal granule cell layer (IGL) of the adult cerebellum, this involves translocation of the cell soma from the external cell layer (EGL) just below the pial surface through the underlying molecule (ML) and Purkinje cell layers (PL). Rakic (193) has carried out extensive electron microscopic analysis of granule cell migration, the results of which are summarized in *Figure 7*. Initially, the post-mitotic granule cell extends two horizontal processes along the surfaces of previously generated granule cell parallel fibers (the horizontal portion of the T-shaped granule cell axon, *Figure 7A*). Subsequently, a third process grows at 90° to these initial processes along Bergmann glial fibers (*Figure 7B*) which extend from Bergmann astroglial cell bodies in the PL through the ML and EGL to the pial surface. The cell soma then begins its inward translocation behind the leading tip of the descending process (*Figure 7C*) to reach its final location in the IGL (*Figure 7D*). As pointed out by Rakic (194), these events involve granule cell interaction with, and guidance along, both neuronal and glial cell processes. Based on immunohistochemical localizations (90,195,196) and *in vitro* cell binding studies (*Table 2*), both NCAM and L1/NgCAM may mediate granule cell–granule

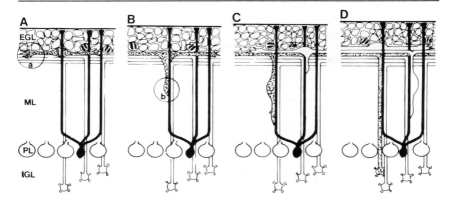

Figure 7. Diagrammatic representation of principle events during cerebellar granule cell migration. The stippled cell represents a migrating granule cell and the darkened cell is a Bergmann glial cell drawn with three processes extending to the pial surface. In (**A**) the stippled granule cell is shown extending two horizontal processes along previously placed granule cell horizontal processes at the inner surface of the EGL. In (**B**) the stippled cell has extended to a third process which is descending along a Bergmann glial cell process. In (**C**) the granule cell soma is shown in the process of translocation through the ML behind the descending process' leading edge. In (**D**) migration of the stippled granule cell soma to the IGL has been completed leaving behind a bifurcating axon. The leading tips of the two horizontal processes (circle a) have ultrastructural features typical of axon growth cones, whilst the ultrastructure of the leading tip of the descending process (circle b) is more similar to that for the leading edge of traditional migrating cells (From ref. 194, with permission.)

cell interactions while NCAM, AMOG, and J1/cytotactin may be involved in granule cell adhesion and leading process extension along Bergmann glial fibers. L1/NgCAM is unlikely to be involved in the latter interactions as at the electron microscope level it is *not* present at sites of granule cell – Bergmann glial cell contact (196) and antibodies to L1 do not inhibit cerebellar astrocyte – neuron adhesion in simple binding assays *in vitro* (38).

Two sorts of *in-vitro* assays have been used to test the functional role of these molecules in granule cell migration. One assay, which involves a system for the microculture of dissociated cerebellar cells, allows the direct visualization of granule cell migration along astroglial processes (197,198). Using this system, Hatten's group has tested the effect of Fab fragments of anti-NCAM and anti-NILE (= L1/NgCAM) antibodies on granule cell – astroglial interactions (199). Neither antibody showed any effect. In contrast a polyspecific antibody prepared against granule cells and absorbed with PC12 cells (which express both NCAM and L1/NgCAM) blocked interaction of granule cells with astrocyte processes. The most prominent granule cell surface antigen recognized by this

Table 2. Inhibition of mouse cerebellar cell adhesion by antibodies to neural adhesion molecules

Antibody	% inhibition of target cell binding to monolayer cultures[a]				Antigen present		Reference
	N–N	N–A	A–A	A–N	Neuron	Astrocyte	
NCAM	40 ± 7	29 ± 7	24 ± 2	28 ± 5	×	×	38
L1	27 ± 4	0 ± 3	−1 ± 2	0 ± 2	×		38
J1	−3 ± 8	47 ± 7	−7 ± 12	38 ± 15		×	86
AMOG	N.T.	26 ± 2	4 ± 2	N.T.		×	90
Control	2 ± 4	−3 ± 2	1 ± 1	−2 ± 3	×	×	86

[a]Neurons (N) and astrocytes (A), purified from early post-natal mouse cerebella, were used as a cell suspension (probe cells) or as a monolayer to which the binding of probe cells was measured. The first cell type mentioned for each pair was used as probe. Numbers are mean values ± SD for several experiments. Without antibodies 50–80% of probe cells bound to cultured monolayers.

absorbed antiserum was a 100 kd glycoprotein, tentatively identified as 'astrotactin'. As expected, absorption of the antibody with normal granule cells removed both the inhibitory activity and recognition of this 100 kd glycoprotein. In contrast, absorption of anti-astrotactin antibodies with granule cells from the mutant mouse, *weaver* (a neurological mutant in which granule cells fail to migrate), removed neither activity. Weaver granule cells express less than 5% the normal granule cell level of this 100 kd antigen. In the dissociated cell microculture system, normal granule cells readily migrate on weaver astrocytic processes while weaver granule cells fail to interact with wild-type astrocytes (200). When examined with high-resolution time-lapse videomicroscopy (199), weaver granule cell filopodia can be seen to repeatedly contact and then withdraw from astroglial processes. In contrast, initial contacts in cultures of normal cerebella are quickly stabilized and persist for hours, frequently with granule cell migration. In the presence of anti-astrotactin antibodies, initial granule cell filopodial contacts with astrocytes do not persist but are quickly withdrawn as in weaver cultures. Thus, the granule cell surface antigen astrotactin may be required for the initial adhesion and recognition of Bergmann glial fibers by granule cells. The reduced level of this glycoprotein on weaver granule cell surfaces may be key to their inability to migrate.

In a second *in-vitro* assay developed by Linder et al. (201), migration of [^{3}H]thymidine-labeled granule cells from the EGL through the ML and PL to the IGL is assayed using cerebellar explants. Small pieces of cerebellar cortex are removed near the onset of granule cell migration and pulse-labeled with [^{3}H]thymidine to label mitotic granule cell nuclei. Immediately after the pulse, nearly all labeled nuclei are found in the EGL

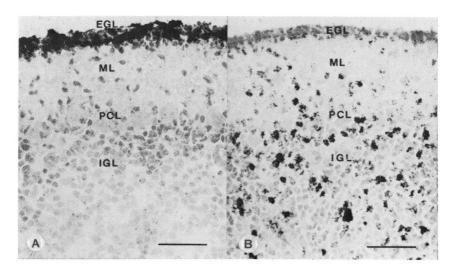

Figure 8. *In-vitro* assay of cerebellar granule cell migration. Autoradiographs of cryostat sections from a piece of post-natal day 10 mouse cerebellum pulse-labeled with [^{3}H]thymidine for 90 min and then maintained for (**A**) 0, or (**B**) 3 days *in vitro*. Sections were counterstained with hematoxylin–eosin to show unlabeled cells. Mitotic granule cells are limited to the EGL (**A**). The inward migration of granule cells *in vitro* occurs with a similar time course as *in vivo*, and can be monitored by the distribution of labeled nuclei between cerebellar layers 2–3 days after pulse-labeling (**B**). (Reproduced with permission from ref. 201.)

(*Figure 8A*). If the explants are then cultured for 3 days, the majority of labeled nuclei are found in the IGL (*Figure 8B*) reflecting the migration of granule cells. Edelman's group has used this system to examine the effects of Fab fragments of antibodies to NCAM, NgCAM, and cytotactin (= J1) (195,202) on granule cell migration in the chick cerebellum, while Schachner's lab has studied the effects of these as well as anti-AMOG antibodies in explants of mouse cerebellum (90,201). Antibodies to NCAM have relatively little effect (8–14% inhibition of migration in both systems) while anti-L1/NgCAM and anti-AMOG antibodies produce a marked inhibition of granule cell migration. Additionally, Chuong *et al.* (195) found anti-cytotactin antibodies inhibited migration in chick cerebellum while in the mouse cerebellar explants anti-J1 was reported to have no effect (90). Although the discrepancy in these results remains to be resolved, it should be pointed out that anti-cytotactin antibodies used by Chuong *et al.* were raised against the larger, embryonic form of J1/cytotactin (200–220 kd) while Schachner's group prepared anti-J1 using J1/cytotactin obtained from adult mouse brain (160 kd). As NCAM is present on both granule cells and Bergmann glial fibers, the small effect of anti-NCAM antibodies provides some assurance that the effects seen with the

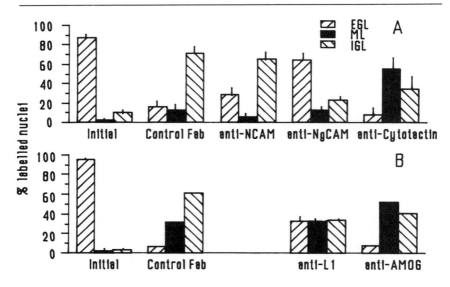

Figure 9. Inhibition of cerebellar granule cell migration *in vitro* by antibodies to neural adhesion molecules. Pieces of (**A**) embryonic day 15 chick or (**B**) postnatal day 10 mouse cerebellum, were pulse-labeled for 60–90 min with [^{3}H]thymidine and then cultured for 3 days in the presence of the indicated antibodies. With the exception of anti-AMOG, which is a monoclonal IgG, all the antibodies were monovalent Fab fragments of polyclonal rabbit antisera. The number of labeled nuclei present in the EGL, ML and IGL (including the PL) were counted microscopically. Values represent the average from several experiments, in which a total of several thousand labeled nuclei were typically counted. Data for chick cerebella (**A**) are from refs 195 and 202; mouse data (**B**) are from refs 90 and 201.

other antibodies are specific. When the distribution of labeled nuclei between the EGL, ML, and IGL (including the PL) is plotted, the inhibition of granule cell migration by L1/NgCAM versus AMOG or cytotactin antibodies can be seen to involve two different steps (*Figure 9*). The effect of anti-L1/NgCAM is largely to inhibit the migration of granule cells out of the EGL with little apparent effect on subsequent migration through the ML. In contrast, when exposed to anti-AMOG or anti-cytotactin antibodies, nearly all labeled granule cells migrated out of the EGL only to accumulate in the deeper ML.

Placing results from both types of studies *in vitro* in the context of the model proposed by Rakic (194) for granule cell migration (*Figure 7*), suggests anti-L1/NgCAM interferes with the initial post-mitotic granule cell process outgrowth along previously positioned parallel fibers, while the astrocyte antigens, cytotactin, and AMOG, as well as granule cell astrotactin are involved in the subsequent elongation of the third granule cell process along Bergmann glial fibers. Thus perhaps at least four separate adhesion/recognition systems may be involved in granule cell migration. Since this involves the recognition and extension of granule cell processes along two different but intersecting pathways (parallel versus Bergmann glial fibers), the granule cell must differentially distribute at least one of these systems along its surface.

3.4 Axon growth

Unlike local circuit neurons such as the cerebellar granule cell whose axon is a direct result of cell migration, the axons of many neurons must extend relatively long distances from the cell soma in order to reach their appropriate target fields. Outgrowth is typically highly ordered, producing an initial connectivity which is very similar to that observed in the adult nervous system (203,204). The growth cone is thought to be the site of axon guidance and elongation during this process (see ref. 205 for a recent review). Filopodial extensions from the growth cone probe the local environment and upon making stable contacts act to redirect further neurite extension in their direction. Mechanistically, the process appears to involve an intracellular tug of war between filopodia, as those which contact more adhesive substrates are preferred (206,207; but see also ref. 208). Indeed, the direction of axon elongation can be experimentally manipulated by applying direct mechanical tension on the growth cone with an adhesive micropipette (209). For some axon pathways, such as the developing insect leg, there seem to be gradients of increasing adhesivity which may guide growth cones towards their target fields (210,211).

Initial axon outgrowth occurs along glial endfeet, neuron cell surfaces, and deposits of ECM. Later, neurites grow along the surfaces of previously extended axons forming large bundles or fascicles. Axon fasciculation is clearly an important process in conducting large numbers of axons from neuron cell body pools to target fields in an orderly array. The choice made by axons to bundle together versus branching to grow individually (or in groups) along other nearby surfaces is likely governed in part by the relative adhesivity of the axon bundle versus these competing substrata. In the following sections, experiments will be reviewed which identify laminin as an important substrate for axon outgrowth and implicate NCAM and L1/NgCAM in neurite fasciculation.

3.4.1 Laminin as a neurite outgrowth-promoting factor

Neurite outgrowth *in vitro* requires the presence of soluble neuron survival factors and suitably adhesive culture substrates. Coating dishes with conditioned media factors, purified ECM components such as fibronectin, collagen and laminin (118,119,212,213), or with clearly artificial substrates such as paladium (207) or poly-lysine all promote neurite outgrowth from cultured embryonic neurons. Of these, laminin (and conditioned media factors) promotes the most robust growth for both central (retinal and spinal motor) and peripheral neurons. To directly demonstrate this,

Gunderson (208) plated neurons on type IV collagen-coated surfaces onto which a patterned overlay of laminin had been added. Neurite outgrowth faithfully reproduced the laminin pattern, avoiding the adjacent collagen surface. In contrast, neurites did not distinguish fibronectin from collagen or collagen-coated surfaces containing fibronectin grids. Supporting the idea that laminin may be an important substrate for axon outgrowth, several groups have found laminin, in a complex with heparan sulfate proteoglycan, to be the principal substrate-binding component of several non-neuronal cell-conditioned media which stimulate neurite outgrowth *in vitro* (121,122,214,215). Laminin co-purifies with neurite-outgrowth-promoting activity in these conditioned media and antibodies to laminin can specifically remove the substrate-binding active component. Furthermore, the monoclonal antibody INO (inhibitor of neurite outgrowth) which blocks neurite outgrowth on conditioned media-treated surfaces, recognizes a laminin–heparan sulfate proteoglycan complex present in conditioned media and in basal lamina *in vivo* (214). When the purified components of conditioned media complex were tested separately, only laminin continued to show neurite-outgrowth-promoting activity (216; but see also ref. 217).

Laminin is clearly present along pathways of nerve outgrowth in the developing PNS (218) suggesting it could also provide a substrate for axon growth *in vivo*. Sandrock and Matthew (219) have recently provided direct support for this idea using the INO monoclonal antibody. Sympathetic innervation of the rat iris was removed by a single injection of 6-hydroxydopamine, a treatment which causes degeneration of terminal and preterminal sympathetic axons but allows survival of neuron cell bodies (220). Within 1 week of lesioning, axons could be seen regrowing into the iris, initially along INO-immunoreactive pathways. Injection of hybridoma cells producing the INO antibody into the anterior chamber of the eye 1 week prior to 6-hydroxydopamine treatment inhibited the subsequent regeneration of sympathetic axons. Injection of hybridomas producing control antibodies which stain the iris in a fashion similar to INO had no effect. Thus, a specific interaction of growing axons with the laminin–heparan sulfate proteoglycan complex recognized by INO may be necessary to stimulate their regeneration.

Until recently, laminin has not been thought to be involved in promoting axon growth in the CNS of higher vertebrates as it is largely restricted to vascular and pial basement membranes. Cohen *et al.* (221) have now demonstrated laminin immunoreactivity along the embryonic chick optic nerve pathway during the period of axon outgrowth. Laminin is present in the area of the neuroepithelial endfeet along which retinal ganglion cell axons grow before initial axon outgrowth but then declines within several days after formation of the optic projection. This transient expression of laminin along the optic pathway correlates with the ability of neural retinal cells from early but not late chick embryos to specifically respond to laminin *in vitro* (140,222). The effect of laminin on neurite outgrowth from retinal neurons appears to involve its binding to the fibronectin/laminin receptor identified by CSAT/JG22 monoclonal antibodies (140,221), yet CSAT/JG22-immunoreactive material continues to be expressed on retinal axons after they no longer show a specific response to laminin. Interestingly, age-dependent changes in the profile of α-like subunits in the integrin receptor were seen for neural retinal cells suggesting that a change in the specificity of binding for this receptor complex may underlie the observed developmental loss of laminin-stimulated neurite outgrowth (140). Liesi (223) and others (224) have pointed out that the presence of laminin along axon pathways correlates with the ability of adult axons to regenerate *in vivo*. Thus, laminin is expressed in the optic tract of adult frogs and goldfish which show optic nerve regeneration, but not in animals (e.g. rats or chickens) which do not (223). Given the developmental change in the response of chick retinal neurons to laminin substrata, more direct tests will be needed to define the potential of laminin to stimulate axon regeneration within the CNS.

3.4.2 Axon–axon interactions

A striking aspect of neurites growing *in vitro* is their tendency to form bundles or fascicles of processes. This side-to-side association reflects a preferential adhesion of neurites to each other as compared to the culture dish. This adhesive comparison can readily be shown by culturing neuron explants (e.g. pieces of spinal cord, retina, sensory ganglia, etc.) on surfaces with different adhesive properties. On tissue culture plastic (a surface with relatively low adhesion for neurons) the halo of neurite outgrowth from explants consists of thick bundles; on collagen-coated dishes, a mixture of small and large fascicules are typically seen; while on very adhesive surfaces such as laminin- or lectin-coated dishes, outgrowth consists of a fine meshwork of neurites.

To assay the role of cell surface adhesion molecules in mediating neurite–neurite interaction, explants cultured on a moderately adhesive surface can be treated with monovalent Fabs of specific antibodies and the thickness of axon bundles measured. In some of their earliest studies of NCAM-mediated adhesion, Rutishauser *et al.* (225) used this assay to demonstrate that anti-NCAM Fabs caused a defasciculation of neurite outgrowth from chick dorsal root ganglia (DRG), (*Figure 10*). Control antibodies raised against chick embryo fibroblasts bound to neurites to a similar degree as anti-NCAM antibodies but had no effect on neurite bundling. Subsequently, similar results have been obtained using several different preparations of anti-NCAM (226–228).

Unlike NCAM, which is present on both neuron cell bodies and their processes, L1/NgCAM is preferentially expressed on axons. In tissue sections, anti-L1/NgCAM strongly stains axon tracts (229–232), suggesting that, in addition to NCAM, L1/NgCAM-mediated

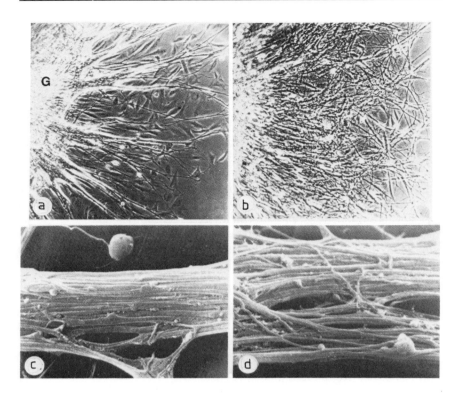

Figure 10. (a) and (b) Phase-contrast micrograph of neurite outgrowth from embryonic day 10 chick thoracic ganglia in the presence of (a) control Fabs or (b) anti-NCAM Fabs (×82). (c) and (d) Scanning electron micrographs of large diameter neurite bundle from (c) a control or (d) anti-NCAM-treated explant (×5000). Anti-NCAM produces a tangled pattern of outgrowth composed predominantly of fine neurite fascicules. Large neurite bundles present in anti-NCAM-treated cultures are less tightly packed than in control explants. (Reproduced with permission from ref. 225.)

neurite–neurite adhesion may also be important for axon bundling. When tested *in vitro*, anti-L1/NgCAM Fabs have been shown to inhibit fasciculation of embryonic chick DRG (202), rat brain (228), and chick retinal (233) axons. Interestingly, under the conditions used in these experiments, anti-NCAM antibodies failed to block axon bundling for DRG and embryonic day 14 (E14) brain neurons (202,228) but either anti-L1/NgCAM or anti-NCAM antibodies could block fasciculation in the retinal explant assay (226,233). In contrast, Stallcup and Beasley (228) found that anti-NCAM but not anti-L1/NgCAM Fabs blocked fasciculation in cultures of post-natal day 5 (P5) rat cerebellar neurons. In their study of neurite bundling in cultures of E14 brain and P5 cerebellum, Stallcup and Beasley (228) used immunoprecipitation to demonstrate the presence and compare the form of NCAM and L1/NgCAM. Both types of cultures contained both adhesion molecules, but NCAM present on E14 brain neurons was heavily polysialylated compared to that on P5 cerebellar cells. Since PSA associated with NCAM reduces the adhesion of reconstituted NCAM-lipid vesicles (see Section 2.1.2), the sialylation of NCAM on E14 brain neurites presumably decreased the influence of NCAM-mediated adhesion in producing neurite fasciculation.

Rutishauser et al. (227) have made a direct test of the influence of NCAM polysialic acid on neurite fasciculation by culturing embryonic chick DRG in the presence of an endoneuraminidase (endo-N) which selectively degrades the unusual α-2,8-linked homopolymers of sialic acid present on NCAM (234). Endo-N treatment of E7 lumbar DRG cultures caused a large reduction in the PSA content of NCAM and produced a marked increase in the degree of neurite bundling, an effect which could be reversed by anti-NCAM antibodies. In contrast, endo-N had little effect on fasciculation of cultured E9 thoracic DRG. At this later stage of development, NCAM in thoracic ganglion contains relatively low levels of PSA and outgrowth from both control and endo-N-treated E9 thoracic DRG explants had thick neurite bundles similar to those seen in endo-N-treated E7 lumbar DRG cultures. More recently this group (235) has reported results which initially appear to be in contradiction to these. Unlike most explants which produce fine neurite outgrowth on laminin, Rutishauser's group found that spinal cord neurons, which contain heavily sialylated NCAM, produce thick axon fascicles on laminin substrates. In contrast to their prior results with E7 DRG explants, treatment of spinal cord cultures with endo-N caused a *de*fasciculation of neurite bundles. To explain this seeming contradiction, these workers suggest that PSA can affect not only NCAM–NCAM adhesion, but due to its large radius of hydration, PSA may sterically prevent the interaction of other cell surface receptors with their respective cell surface- or substrate-bound ligands. To support their argument, this group also demonstrated that high levels of NCAM PSA could inhibit other cell surface contact-mediated interactions not directly involving NCAM adhesion. Thus, axon fasciculation is regulated by a complex set of neurite–neurite and neurite–substratum interactions involving several adhesion molecules. The apparent contribution to axon fasciculation by a given molecule depends on the presence or absence of other competing or reinforcing adhesion systems.

Studies of neurite fasciculation, such as those described above provide important information about the role of competing neurite–neurite and neurite–substratum adhesion in maintaining or disrupting axon bundles. But, as neurite outgrowth in these experiments occurs along both axon surfaces and the culture dish, the role of specific adhesion molecules in neurite extension along axon surfaces is not easily assayed. Chang et al.

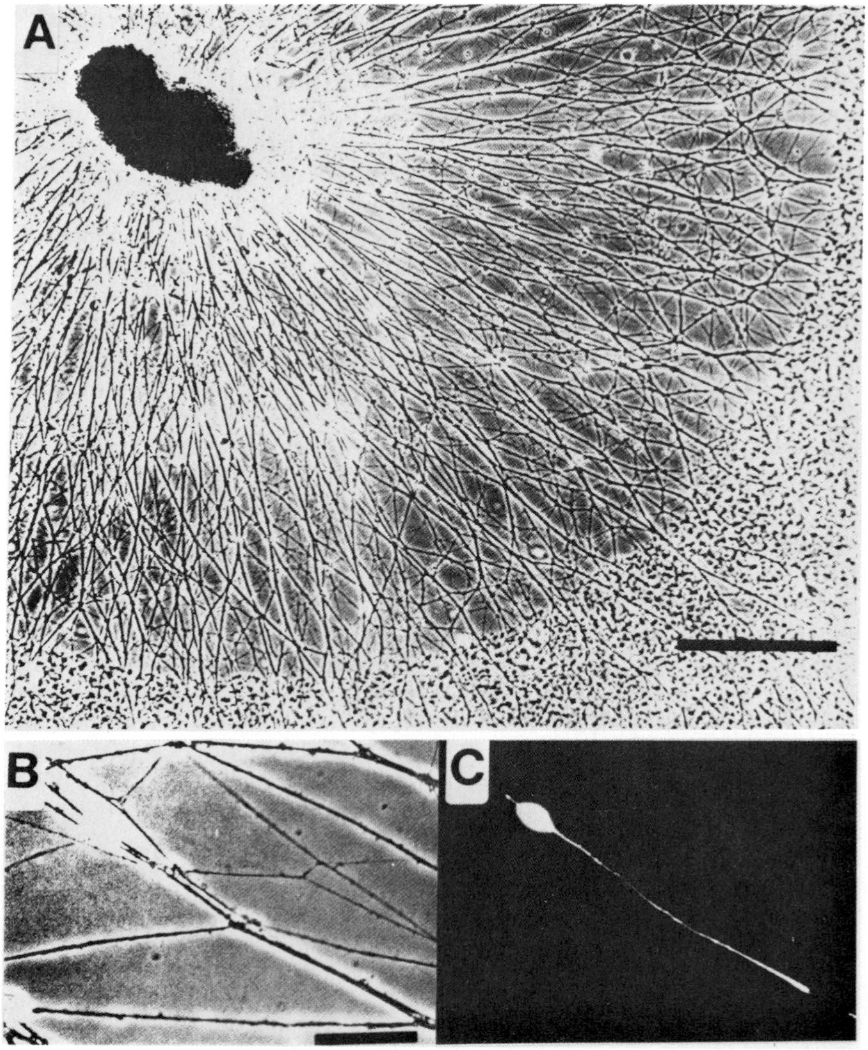

Figure 11. *In-vitro* assay of neurite outgrowth along axon surfaces. (**A**) Photomicrograph of sympathetic ganglia explant prepared as described in the text for use as a neurite outgrowth substrate. A fine meshwork of axons has extended from the explant (upper left corner) to reach the annulus of fixed cells (visible along the right and bottom edges of panel). (**B**) Higher magnification phase-contrast and (**C**) fluorescence photomicrographs of axonal substrate with an attached, fluorescently labeled test neuron which has extended a neurite along an explant process. The bar in (**A**) represents 500 μm and that in (**B**), 50 μm. (From ref. 226 with permission.)

(226) have recently devised modifications of more traditional fasciculation assays which allow direct examination of the importance of specific axon surface molecules in stimulating neurite outgrowth. In this system, a neuron explant is placed in the center of a laminin-coated coverslip which contains an annulus of methanol/acetone-fixed brain cells. After several days in culture a fine radial outgrowth of neurites is produced along the laminin substrate and extends to the ring of fixed cell. Anti-laminin antibodies are then added to the culture along with the freshly dissociated fluorescently labeled test neurons. Antibodies to laminin cause the previously extended axon processes to release from the laminin culture surface, laminin-independent adhesion of distal axon surfaces to the annulus of fixed cells maintains their position. Test neurons readily bind to the explant outgrowth and extend neurites along these axon surfaces. By neutralizing neurite − substratum interactions, anti-laminin antibodies allow examination of neurite outgrowth along explant axons in the absence of competing neurite − substrate adhesion (*Figure 11*). Chang et al. (226) used this assay to test the effect of Fab' fragments of antibodies to NCAM, L1/NgCAM, and a third, less well characterized neural adhesion molecule, F11, on neurite outgrowth from chick sympathetic neurons along axons of sympathetic ganglia explants. Although all three antigens could be demonstrated on the surface of explant axons by immunofluorescence staining, only antibodies of L1/NgCAM and the F11 antigen reduced the length of neurite outgrowth from added test neurons. Anti-NCAM Fabs were without effect. None of the antibodies affected outgrowth on laminin surfaces suggesting the effects of anti-L1/NgCAM and anti-F11 did not reflect a general inhibition of neurite growth. Although no data were presented as to the ability of anti-NCAM antibodies to disrupt neurite fasciculation of E9 sympathetic explants, if they do, these results would suggest that NCAM might produce functionally important adhesion along the neurite shaft but provide little motivation for growth cone extension. As the effects of anti-L1/NgCAM and anti-F11 antibodies were partial (producing neurite outgrowth 60 − 80% that of control) it would also be interesting to test the effect of anti-NCAM antibodies in the presence of one or both of the other antibodies.

3.5 Cell adhesion molecules and neural specificity

The mechanisms by which the very specific neural connections evident in the mature nervous system arise has long occupied the interest of biologists. Perhaps the most influential hypothesis regarding this topic has been Sperry's chemoaffinity theory (236) which proposes that gradients of chemical markers are used to match pre- and post-synaptic cells. In the chick retinotectal system, corresponding gradients of a cell surface antigen, TOP (for toponymic), occur on the surface of retinal and tectal cells (237). The retina contains a ventral-to-dorsal gradient of

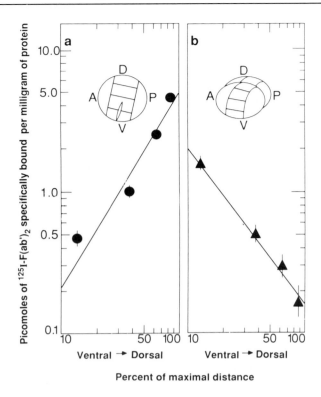

Figure 12. TOP antigen gradients in (**a**) embryonic day 5 chick retina and (**b**) tectum. Strips of retina and optic tectum (1 × 2 mm) were cut along the dorsal–ventral (D–V) axis as shown and TOP antigen was measured using a solid-phase radioimmunoassay. A quantitatively similar antigen gradient is seen in both the retina and tectum but with reversed dorsal–ventral polarity. A and P refer to the anterior and posterior axis. (Reproduced with permission from ref. 237.)

increasing TOP antigen concentration, while in the tectum, the gradient is reversed (*Figure 12*). The pattern of this gradient parallels the retinotectal projection, dorsal areas of the retina project to ventral tectum and ventral retina to dorsal tectum. Since the tectal gradient is present before arrival of retinal axons, it is not simply a reflection of retinal axon TOP antigen. Although little is known about the nature or functional activity of the TOP antigen, this intriguing distribution suggests it could serve to order a dorsal–ventral topographic projection of retinal axons onto the tectum.

In addition to positional adhesive labels, the orderly formation and refinement of neural connections is likely to involve more general adhesive

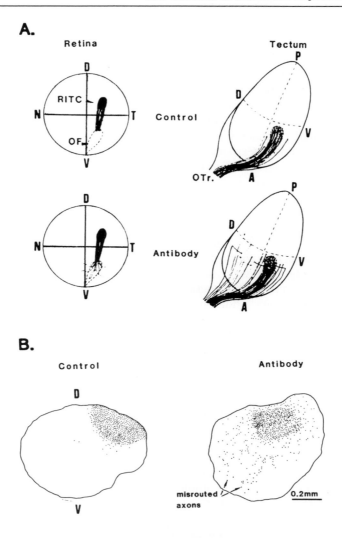

Figure 13. Intra-occular injection of anti-NCAM antibodies disrupts the orderly fasciculation of growing retinal axons. Anti-NCAM Fabs were injected on embryonic day 4. The route of axons originating from a distinct site in the retina were labeled 5 days later by the localized uptake and anterograde axonal transport of rhodamine isothiocyanate (RITC) from a RITC crystal placed on the retina. (**A**) Drawing of RITC-labeled axons in the retina and optic tectum. In anti-NCAM-treated embryos, retinal axons become disordered as they leave the retina at the optic fissure (OF). At the optic tectum, some misrouted axons make abrupt, right angle turns to partially correct their position. (**B**) Cross-sections through the optic nerve showing the decreased coherence of RITC-labeled axons from an anti-NCAM injected eye. D, dorsal; V, ventral; N, nasal; T, temporal; P, posterior; A, anterior; OTr, optic tract. (From ref. 241, with permission.)

interactions as well (238–240). For example, for axons with topographic projections, position-independent adhesion systems involved in axon fasciculation could greatly simplify the task required to reach correct synaptic sites by maintaining neighbor relationships. Thanos et al. (241) have provided a striking demonstration of this by perturbing axon–axon interactions of retinal ganglion cells in the eye using anti-NCAM antibodies. The clustered topographic projection of retinal axons can be demonstrated by placing a crystal of rhodamine isothiocyanate on the retina to label axons arising from that point. The distribution of these axons in the tectum as well as the pathway they follow to reach their target can be easily observed in the fluorescent microscope. As shown in *Figure 13*, chick retinal axons arising in a given region of the retina arrive at the tectum as a bundle and terminate in close relationship to one another. Injection of anti-NCAM Fabs into the eyecup at the time of retinal axon outgrowth results in a significant number of axons which lose their initial orderly relationship with other nearby fibers as they leave the eye. This decreased coherence was quite obvious in optic nerve cross-sections and resulted in inappropriate tectal projections by approximately 15% of labeled retinal axons. Addition of purified NCAM to the antibodies before injection neutralized their effect. Since the injected antibodies did not penetrate beyond the optic fissure, the errors in tectal projection were unlikely the result from direct antibody effects at synaptic sites but rather were the manifestation of a disordered bundling of axons leaving the eye. Experiments by Fraser et al. (242,243) suggest that, in addition to maintaining axon bundles, position-independent NCAM-mediated adhesion may also be important in later activity-dependent refinements of initial retinotectal projections.

3.6 Nerve – muscle adhesion

Owing to its relative simplicity and accessibility, the neuromuscular system has been popular for studies of neuron–target interactions. Spinal motor axons grow along stereotyped pathways to reach their peripheral targets and upon contact with embryonic myotubes quickly form synapses. While the chemical signals which are involved in this process are poorly understood, most would agree that a cell surface adhesion/recognition system is involved in promoting nerve–muscle contact. At least two neural adhesion molecules, NCAM and N-cadherin are expressed by embryonic muscle (75,244–246), and experiments *in vitro* suggest both may be involved in neurite extension along myotube surfaces. Rutishauser et al. (247) co-cultured chick spinal cord neurons with embryonic myotubes in the presence of anti-NCAM Fabs or control Fabs which also bound to both nerve and muscle surfaces. Anti-NCAM, but not control Fabs, prevented the normal, extensive neurite outgrowth along myotube surfaces. Antibodies did not prevent neurite growth *per se*, but, instead of being predominantly along myotubes, outgrowth was shifted to culture surfaces between myotubes. Bixby et al. (248) have obtained similar results using co-cultures of chick ciliary ganglia neurons and embryonic myotubes. Antibodies of NCAM and N-cadherin decreased neurite contact with myotubes, causing proportionately greater contact with fibroblasts and ECM present between myotubes. By using the JG22 monoclonal antibody to selectively block the latter neurite interactions, this group clearly demonstrated that both NCAM and N-cadherin can promote nerve−muscle contact. The separate inhibition of neurite−myotube contact by anti-NCAM and anti-N-cadherin antibodies were additive when used in combination suggesting these two adhesion systems act independently.

Although detailed study of the distribution of muscle N-cadherin in vivo has not been reported, a great deal of information about the distribution and regulation of NCAM in muscle has recently been obtained. The results of these studies support the idea that NCAM may be involved in promoting initial nerve−muscle contacts. During development in vivo, NCAM is present on both motor axons and myotube surfaces at the time of initial nerve−muscle contact (244,245). NCAM occurs along the entire length of myotubes suggesting it is unlikely to determine where along the myotube initial neurite contacts or synapses will form. In chick limb muscles, the levels of muscle NCAM dramatically increase coincident with extensive motor axon ramification within muscle masses (246) (*Figure 14*). This increase is not dependent on the arrival of motor axons as it also occurs in aneural limbs produced by surgically removing spinal cord segments prior to axon outgrowth (246). The temporal relationship between increased levels of muscle NCAM and intramuscular axon sprouting and growth suggests a functional role for NCAM-mediated adhesion in this process. Results from more recent studies from Landmesser's group supports this idea (249). Fab fragments of anti-NCAM and anti-L1/NgCAM antibodies were injected into the developing chick limb bud at the time of extensive intramuscular axon growth and their effects visualized by staining whole mount preparations of individual muscles with antibodies to neurofilament antigens. A dramatic increase in the number of side branches leaving the main nerve trunk was seen in embryos treated with anti-L1/NgCAM Fabs and a *decreased* number of side branches in embryos treated with anti-NCAM. The results using anti-L1/NgCAM parallel the effects of these antibodies on neurite fasciculation *in vitro* (202,228,233), causing an apparent increase in neurite interaction with non-neural substrates analogous to defasciculation. Unlike L1/NgCAM, which is absent from embryonic muscle fiber surfaces (250), NCAM is present both on neurite surfaces and on the major alternative substrate for neurite growth in these experiments, the muscle fiber surface. Thus, anti-NCAM antibodies could inhibit both neurite−neurite and neurite−substrate (myotube) interactions. Dependent upon the relative contribution of NCAM-mediated adhesion to the two processes,

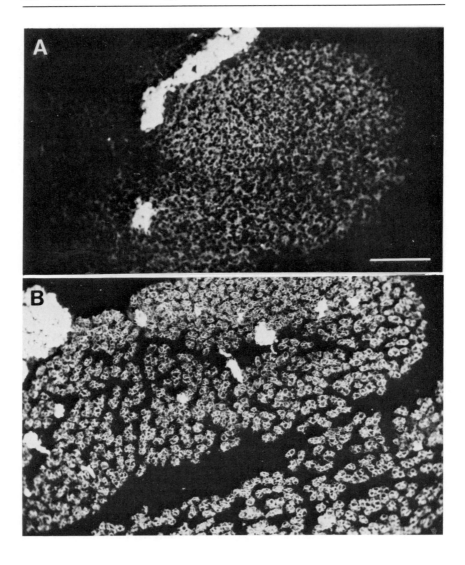

Figure 14. Cross-sections of the iliofibularis and posterior iliotibialis muscles stained with a monoclonal anti-NCAM antibody and fluorescein-conjugated second antibody. (**A**) Section from a stage 28–29 chick embryo, prior to the intramuscular growth of nerve fibers. Primary muscle nerve branches have left the sciatic nerve and are brightly stained. Myotubes are present in the muscles but contain much lower levels of NCAM. (**B**) Cross-section from a level comparable to that in (**A**) but from a stage 33 embryo. Extensive intramuscular nerve branches are now present, both nerves and myotubes contain high levels of NCAM. Bar represents 100 μm. (Photomicrographs kindly provided by L.Dahm and L.Landmesser.)

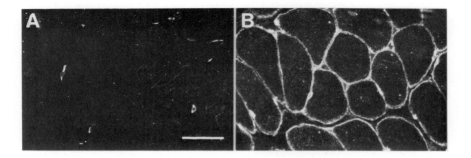

Figure 15. Cross-sections of (**A**) normal or (**B**) 7-day denervated adult chicken muscle (flexor digitorum profundus) stained with anti-NCAM antibodies and fluorescein-conjugated second antibody. Muscle denervation induces the accumulation of NCAM throughout the muscle fiber surface. Structures stained by anti-NCAM in panel (**A**) are muscle satellite cells. The bar represents 50 μm.

anti-NCAM could either increase or decrease side branch information (i.e. defasciculation). The observed *reduction* in side branch formation by anti-NCAM antibodies suggests that *in vivo*, functional NCAM-mediated neurite–myotube adhesion occurs.

Shortly after birth (hatching), levels of extra-synaptic muscle fiber NCAM are dramatically reduced (244,246,251). NCAM persists at the neuromuscular junction (244,251,252) where it is associated with both pre- and post-synaptic structures (244). Endplate-associated NCAM may serve to bind the nerve terminal Schwann cell and myelin-free terminal axon, but is unlikely to mediate adhesion between the nerve terminal and post-synaptic gutter since muscle NCAM is concentrated at the depths of post-synaptic folds. The crest of folds, which have high concentrations of acetylcholine receptors, show little NCAM immunoreactivity (244). Expression of NCAM in extra-junctional areas of the adult muscle fiber is regulated by nerve-evoked electrical or contractile activity (252–254). Denervation or paralysis of adult muscle causes a rapid accumulation of NCAM, (*Figure 15*), while re-innervation of previously denervated muscle produces the opposite effect. This activity-dependent regulation appears to occur at the RNA level as denervation of rodent muscle causes a 30- to 50-fold increase in the level of muscle NCAM RNA (8). Neural regulation of muscle NCAM expression, suggests NCAM-mediated adhesion could be part of a feed-back loop of nerve–muscle interaction which contributes together with other similarly regulated muscle cell surface and soluble factors (255) to the overall regulation of synaptogenesis. Hopefully, future *in-vivo* studies, such as those being conducted by Landmesser's group will begin to provide a more rigorous test of these ideas.

Walsh's group (256) has recently added a new and intriguing detail to this description of the possible involvement of NCAM in nerve-muscle recognition. They find that embryonic muscle NCAM RNA contains a muscle-specific 108 nucleotide sequence located between exons 12 and 13 in the external domain of NCAM. While the importance of the predicted novel 38 amino acid insert remains to be identified, its location in the extracellular domain suggests it could modulate NCAM adhesion, perhaps allowing the discrimination of nerve versus muscle NCAM by growing motor axons. Given the large size of the NCAM gene (at least 100 kb in the mouse; ref. 10), this result, coupled with those from Akeson's group (67; Section 2.4) describing a 10 amino acid insertion in the fourth immunoglobulin domain of NCAM in a subset of neurons, raises the possibility that other populations of NCAM-positive cells could also use alternative RNA splicing to produce extensive heterogeneity in the extracellular domain of NCAM.

4. Conclusions

The list of characterized neural adhesion molecules has grown rapidly during the past several years. During this time it has become increasingly clear that adhesive interactions represent the summation of distinct competing and reinforcing molecular interactions. An understanding of the impact of individual interactions, as well as identification of new ones, will require functional studies in which several adhesion systems are manipulated simultaneously. From an evolutionary viewpoint, it is interesting to note that several of these adhesion molecules— NCAM/L1-NgCAM/MAG, N-cadherin, laminin/S-laminin, and integrin— are actually members of multigene families. The cytoadhesin family provides an exciting example of how molecular recombination of related sequences and subunits could produce very different adhesive specificities. Given recent advances in other areas of molecular biology, such relationships are perhaps not unexpected. Overlain on this apparent use of molecular evolution to provide diversity, alternative RNA splicing produces multiple, and sometimes functionally distinct, adhesion molecules from single genes (e.g. fibronectin, NCAM, and MAG). Sequence analysis for other adhesion molecules will undoubtedly provide additional examples of the impact of molecular genetic processes on cell adhesion.

Macroscopic cell adhesion appears to rely on a high prevalence of relatively low affinity adhesion molecules. This relationship may be important to allow readily reversible cell–cell interactions as well as providing a mechanism to grade adhesion strength by modulating CAM prevalence (30). The ability of cells to modulate the level and pattern of expression of multiple adhesion factors is likely an important mechanism for producing both cell type and developmental changes in adhesive specificities. The importance of undertaking molecular studies of the mechanisms involved in regulating the expression of these adhesion systems is clear.

Future studies of cell–cell interaction will undoubtedly identify both additional adhesion molecules and also a new class of cell surface molecules; factors which influence cell–cell contact not by directly mediating cell binding, but rather by providing surface labels for cell recognition (e.g. 257,258). This promises to be an exciting era for students of cell surface interactions in neural development.

5. Acknowledgements

I would like to thank Drs Lynn Landmesser, Mary Hatten, and Urs Rutishauser for sharing their results prior to publication, Rob Pilarski for help in preparing the manuscript and Dr Andy Moiseff and Mary Jane Spring for help with illustrations. The author is supported by a fellowship from the Alfred P.Sloan Foundation and by a grant from the NIH (NS25264).

6. References

1. Wilson,H.V. (1907) On some phenomena of coalescence and regeneration in sponges. *J. Exp. Zool.*, **5**, 245.
2. Brackenbury,R., Thiery,J.P., Rutishauser,U. and Edelman,G.M. (1977) Adhesion among neural cells of the chick embryo I. An immunological assay for molecules involved in cell–cell binding. *J. Biol. Chem.*, **252**, 6835.
3. Thiery,J.P., Brackenbury,R., Rutishauser,U. and Edelman,G.M. (1977) Adhesion among neural cells of the chick embryo II. Purification and characterization of a cell adhesion molecule from neural retina. *J. Biol. Chem.*, **252**, 6841.
4. Lemmon,V., Staros,E.B., Perry,H.E. and Gottlieb,D.I. (1982) A monoclonal antibody which binds to the surface of chick brain cells and myotubes: cell selectivity and properties of the antigen. *Dev. Brain Res.*, **3**, 349.
5. Hirn,M., Pierres,M., Deagostini-Bazin,H., Hirsch,M. and Goridis,C. (1981) Monoclonal antibody against cell surface glycoprotein of neurons. *Brain Res.*, **214**, 433.
6. Rasmussen,S., Ramlau,J., Axelsen,N.H. and Bock,E. (1982) Purification of the synaptic membrane glycoprotein D2 from rat brain. *Scand. J. Immunol.*, **15**, 179.
7. Crossin,K.L., Chuong,C.M. and Edelman,G.M. (1985) Expression sequences of cell adhesion molecules. *Proc. Natl. Acad. Sci. USA*, **82**, 6942.
8. Covault,J., Merlie,J.P., Goridis,C. and Sanes,J.R. (1986) Molecular forms of N-CAM and its RNA in developing and denervated skeletal muscle. *J. Cell Biol.*, **102**, 731.
9. Cunningham,B.A., Hemperly,J.J., Murray,B.A., Prediger,E.A., Brackenbury,R. and Edelman,G.M. (1987) Neural cell adhesion molecule: structure, immunoglobulin-like domains, cell surface modulation, and alternative RNA splicing. *Science*, **236**, 799.
10. Gennarini,G., Hirsch,M.R., He,H.T., Hirn,M., Finne,J. and Goridis,C. (1986) Differential expression of mouse neural cell-adhesion molecule (N-CAM) mRNA species during brain development and in neural cell. *J. Neurosci.*, **6**, 1983.
11. Murray,B.A., Hemperly,J.J., Prediger,E.A., Edelman,G.M. and Cunningham,B.A. (1986) Alternatively spliced mRNAs code for different polypeptide chains of the chicken neural cell adhesion molecule (N-CAM). *J. Cell Biol.*, **102**, 189.
12. Owens,G.C., Edelman,G.M. and Cunningham,B.A. (1987) Organization of the neural cell adhesion molecule (N-CAM) gene: alternative exon usage as the basis for different membrane-associated domains. *Proc. Natl. Acad. Sci. USA*, **84**, 294.

13. Barthels,D., Santoni,M.J., Wille,W., Ruppert,C., Chaix,J.C., Hirsch,M.R., Fontecilla-Camps,J.C. and Goridis,C. (1987) Isolation and nucleotide sequence of mouse NCAM cDNA that codes for a Mr 79 000 polypeptide without a membrane-spanning region. *EMBO J.*, **6**, 907.
14. Hemperly,J.J., Edelman,G.M. and Cunningham,B.A. (1986) cDNA clones of the neural cell adhesion molecule (N-CAM) lacking a membrane-spanning region consistent with evidence for membrane attachment via a phosphatidylinositol intermediate. *Proc. Natl. Acad. Sci. USA*, **83**, 9822.
15. He,H.T., Barbet,J., Chaix,J.C. and Goridis,C. (1986) Phosphatidylinositol is involved in the membrane attachment of NCAM-120, the smallest component of the neural cell adhesion molecule. *EMBO J.*, **5**, 2489.
16. Finne,J., Finne,U., Deagostini-Bazin,H. and Goridis,C. (1983) Occurrence of 2–8 linked polysialosyl units in a neural cell adhesion molecule. *Biochem. Biophys. Res. Commun.*, **112**, 482.
17. Hoffman,S., Sorkin,B.C., White,P.C., Brackenbury,R., Mailhammer,R., Rutishauser,U., Cunningham,B.A. and Edelman,G.M. (1982) Chemical characterization of a neural cell adhesion molecule purified from embryonic brain membranes. *J. Biol. Chem.*, **257**, 7720.
18. Rothbard,J.R., Brackenbury,R., Cunningham,B.A. and Edelman,G.M. (1982) Differences in the carboydrate structures of neural cell-adhesion molecules from adult and embryonic chicken brains. *J. Biol. Chem.*, **257**, 11064.
19. Sunshine,J., Balak,K., Rutishauser,U. and Jacobson,M. (1987) Changes in neural cell adhesion molecule (NCAM) structure during vertebrate neural development. *Proc. Natl. Acad. Sci. USA*, **84**, 5986.
20. Pollerberg,E.G., Sadoul,R., Goridis,C. and Schachner,M. (1985) Selective expression of the 180-kd component of the neural cell adhesion molecule N-CAM during development. *J. Cell Biol.*, **101**, 1921.
21. Chuong,C.M. and Edelman,G.M. (1984) Alterations in neural cell adhesion molecules during development of different regions of the nervous system. *J. Neurosci.*, **4**, 2354.
22. Schlosshauer,B., Schwarz,U. and Rutishauser,U. (1984) Topological distribution of different forms of neural cell adhesion molecule in the developing chick visual system. *Nature*, **310**, 141.
23. Rutishauser,U., Hoffman,S. and Edelman,G.M. (1982) Binding properties of a cell adhesion molecule from neural tissue. *Proc. Natl. Acad. Sci. USA*, **79**, 685.
24. Cunningham,B.A., Hoffman,S., Rutishauser,U., Hemperly,J.J. and Edelman,G.M. (1983) Molecular topography of the neural cell adhesion molecule N-CAM: surface orientation and location of sialic acid-rich and binding regions. *Proc. Natl. Acad. Sci. USA*, **80**, 3116.
25. Cole,G. and Glaser,L. (1986) A heparin-binding domain from N-CAM is involved in neural cell–substratum adhesion. *J. Cell Biol.*, **102**, 403.
26. Cole,G.J., Schubert,D. and Glaser,L. (1985) Cell–substratum adhesion in chick neural retina depends upon protein–heparan sulfate interactions. *J. Cell Biol.*, **100**, 1192.
27. Cole,G., Loewy,A., Cross,N.V., Akeson,R. and Glaser,L. (1986) Topographic localization of the heparin-binding domain of the neural cell adhesion molecule N-CAM. *J. Cell Biol.*, **103**, 1739.
28. Cole,G., Loewy,A. and Glaser,L. (1986) Neuronal cell–cell adhesion depends on interactions of N-CAM with heparin-like molecules. *Nature*, **320**, 445.
29. Crossin,K.L., Edelman,G.M. and Cunningham,B.A. (1984) Mapping of three carbohydrate attachment sites in embryonic and adult forms of the neural cell adhesion molecule. *J. Cell Biol.*, **99**, 1848.
30. Hoffman,S. and Edelman,G.M. (1983) Kinetics of homophilic binding by embryonic and adult forms of the neural cell adhesion molecule. *Proc. Natl. Acad. Sci. USA*, **80**, 5762.
31. Sadoul,R., Hirn,M., Deagostini-Bazin,H., Rougon,G. and Goridis,C. (1983) Adult and embryonic mouse neural cell adhesion moelcules have different binding properties. *Nature*, **304**, 347.
32. Pollerberg,G.E., Schachner,M. and Davoust,J. (1986) Differentiation state-dependent surface mobilities of two forms of the neural cell adhesion molecule. *Nature*, **324**, 462.
33. Hall,A.K. and Rutishauser,U. (1987) Visualization of neural cell adhesion molecule by electron microscopy. *J. Cell Biol.*, **104**, 1579.

34. Grumet,M., Rutishauser,U. and Edelman,G.M. (1983) Neuron-glia adhesion is inhibited by antibodies to neural determinants. *Science,* **222**, 60.
35. Goridis,C., Deagostini-Bazin,H., Hirn,M., Hirsch,M.R.., Rougon,G., Sadoul,R., Langley,O.K., Gombos,G. and Finne,J. (1983) Neural surface-antigens during nervous-system development. *Cold Spring Harbor Symp. Quant. Biol.,* **48**, 527.
36. Noble,M., Albrechtesen,M., Moller,C., Lyles,J., Bock,E., Goridis,C., Watanabe,M. and Rutishauser,U. (1985) Glial cells express N-CAM/D2-CAM-like polypeptides *in vitro. Nature,* **316**, 725.
37. Silver,J. and Rutishauser,U. (1984) Guidance of optic axons *in vivo* by a performed adhesive pathway on neuroepithelial endfeet. *Dev. Biol.,* **106**, 485.
38. Keilhauer,G., Faissner,A. and Schachner,M. (1985) Differential inhibition of neurone-neurone, neuron-astrocyte and astrocyte-astrocyte adhesion by L1, L2, and N-CAM antibodies. *Nature,* **316**, 728.
39. Grumet,M. and Edelman,G.M. (1984) Heterotypic binding between neuronal membrane vesicles and glial cells is mediated by a specific cell adhesion molecule. *J. Cell Biol.,* **98**, 1746.
40. Grumet,M., Hoffman,S. and Edelman,G.M. (1984) Two antigenically related neuronal cell adhesion molecules of different specificities mediate neuron-neuron and neuron-glia adhesion. *Proc. Natl. Acad. Sci. USA,* **81**, 267.
41. Grumet,M., Hoffman,S., Chuong,C.M. and Edelman,G.M. (1984) Polypeptide components and binding functions and neuron-glia cell adhesion molecules. *Proc. Natl. Acad. Sci. USA,* **81**, 7989.
42. Kruse,J., Mailhammer,R., Wernecke,H., Faissner,A., Sommer,I., Goridis,C. and Schachner,M. (1984) Neural cell adhesion molecules and myelin-associated glycoprotein share a common carbohydrate moiety recognized by monoclonal antibodies L2 and HNK-1. *Nature,* **311**, 153.
43. Rathjen,F.G. and Schachner,M. (1984) Immunocytological and biochemical characterization of a new neuronal cell surface component (L1 antigen) which is involved in cell adhesion. *EMBO J.,* **3**, 1.
44. Rathjen,F.G. and Rutishauser,U. (1984) Comparison of two cell surface molecules involved in neural cell adhesion. *EMBO J.,* **3**, 461.
45. Faissner,A., Kruse,J., Nieke,J. and Schachner,M. (1984) Expression of neural cell-adhesion molecule L1 during development, in neurological mutants and in the peripheral nervous-system. *Dev. Brain Res.,* **15**, 69.
46. Salton,S.R.J., Richter-Landsberg,C., Greene,L.A. and Shelanski,M.L. (1983) Nerve growth factor-inducible large external (NILE) glycoprotein: studies of a central and peripheral neuronal marker. *J. Neurosci.,* **3**, 441.
47. Bock,E., Richter-Landsberg,C., Faissner,A. and Schachner,M. (1985) Demonstration of immunochemical identity between the nerve growth factor-inducible large external (NILE) glycoprotein and the cell adhesion molecule L1. *EMBO J.,* **4**, 2765.
48. Friedlander,D.R., Grumet,M. and Edelman,G.M. (1986) Nerve growth factor enhances expression of neuron-glia cell adhesion molecule in PC12 cells. *J. Cell Biol.,* **102**, 413.
49. McGuire,J.C., Greene,L.A. and Furano,A.V. (1978) NGF stimulates incorporation of fucose or glucosamine into an external glycoprotein in cultured rat PC12 pheochromocytoma cells. *Cell,* **15**, 357.
50. Grumet M. and Edelman,G.M. (1988) Neuron-glia cell adhesion molecule interacts with neurons and astroglia via different binding mechanisms. *J. Cell Biol.,* **106**, 487.
51. Laursen,R.A., Samiullah,M. and Lees,M.B. (1984) The structure of bovine brain myelin proteolipid and its organization in myelin. *Proc. Natl. Acad. Sci. USA,* **81**, 2912.
52. Duncan,I.D., Hammang,J.P. and Trapp,B.D. (1987) Abnormal compact myelin in the myelin-deficient rat: Absence of proteolipid protein correlates with a defect in the intraperiod line. *Proc. Natl. Acad. Sci. USA,* **84**, 6287.
53. Lemke,G. and Axel,R. (1985) Isolation and sequence of a cDNA encoding the major structural protein of peripheral myelin. *Cell,* **40**, 501.
54. Trapp,B.D., Quarles,R.H. and Suzuki,K. (1984) Immunocytochemical studies of quaking mice support a role for the myelin-associated glycoprotein in forming and maintaining the periaxonal space and periaxonal cytoplasmic collar of myelinating schwann cells. *J. Cell Biol.,* **99**, 594.
55. Martini,R. and Schachner,M. (1986) Immunoelectron microscopic localization of neural

cell-adhesion molecule (L1, N-CAM, and MAG) and their shared carbohydrate epitope and myelin basic-protein in developing sciatic-nerve. *J. Cell Biol.*, **103**, 2439.
56. Poltorak,M., Sadoul,R., Keilhauer,G., Landa,C., Fahrig,T. and Schachner,M. (1987) Myelin-associated glycoprotein, a member of the L2/HNK-1 family of neural cell adhesion molecules, is involved in neuron–oligodendrocyte and oligodendrocyte—oligodendrocyte interaction. *J. Cell Biol.*, **105**, 1893.
57. Santoni,M.J., Barthels,D., Barbas,J.A., Hirsch,M.R., Steinmetz,M., Goridis,C. and Wille,W. (1987) Analysis of cDNA clones that code for the transmembrane forms of the mouse neural cell adhesion molecule (NCAM) and are generated by alternative RNA splicing. *Nucleic Acids Res.*, **15**, 8621.
58. Moos,M., Tacke,R., Scherer,H., Teplow,D., Fruh,K. and Schachner,M. (1988) Neural adhesion molecule L1 as a member of the immunoglobulin superfamily with binding domains similar to fibronectin. *Nature*, **334**, 701.
59. Arquint,M., Roder,J., Chia,L., Down,J., Wilkinson,D., Bayley,H., Braun,P. and Dunn,R. (1987) Molecular cloning and primary structure of myelin-associated glycoprotein. *Proc. Natl. Acad. Sci. USA*, **84**, 600.
60. Lai,C., Brow,M.A., Nave,K.A., Noronha,A.B., Quarles,R.H., Bloom,F.E., Milner,R.J. and Sutcliffe,J.G. (1987) Two forms of 1B236/myelin-associated glycoprotein, a cell adhesion molecule for postnatal neural development, are produced by alternative splicing. *Proc. Natl. Acad. Sci. USA*, **84**, 4337.
61. Salzer,J.L., Holmes,W.P. and Colman,D.R. (1987) The amino acid sequences of the myelin-associated glycoproteins: homology to the immunoglobulin gene superfamily. *J. Cell Biol.*, **104**, 957.
62. Simmons,D., Makgoba,M.W. and Seed,B. (1988) ICAM, an adhesion ligand of LFA-1, is homologous to the neural cell adhesion molecule NCAM. *Nature*, **331**, 624.
63. Williams,A.F. and Barclay,A.N. (1988) The immunoglobulin superfamily-domains for cell surface recognition. *Annu. Rev. Immunol.*, **6**, 381.
64. Seeger,M.A., Haffley,L. and Kaufman,T.C. (1988) Characterization of amalgam: A member of the immunoglobulin superfamily from Drosophila. *Cell*, **55**, 589.
65. Harrelson,A.L. and Goddman,C.S. (1988) Growth cone guidance in insects: Fasciclin II is a member of the immunoglobulin superfamily. *Science*, **242**, 700.
66. Cole,G.J. and Akeson,R. (1989) Identification of a heparin binding domain of the neural cell adhesion molecule NCAM using synthetic peptides. *Neuron*, **2**, 1157.
67. Small,S.J., Haines,S.L. and Akeson,R.A. (1988) Polypeptide variation in an NCAM extracellular immunoglobulin-like fold is developmentally regulated through alternative splicing. *Neuron*, **1**, 1007.
68. Brackenbury,R., Rutishauser,U. and Edelman,G.M. (1981) Distinct calcium-independent and calcium-dependent adhesion systems of chicken embryo cells. *Proc. Natl. Acad. Sci. USA*, **78**, 387.
69. Grunwald,G.B., Geller,R.L. and Lilien,J. (1980) Enzymatic dissection of embryonic cell adhesive mechanisms. *J. Cell Biol.*, **85**, 766.
70. Magnani,J.L., Thomas,W.A. and Steinberg,M.S. (1981) Two distinct adhesion mechanisms in embryonic chick neural retina cells. I. A kinetic analysis. *Dev. Biol.*, **81**, 96.
71. Takeichi,M., Ozaki,H.S., Tokunaga,K. and Okada,T.S. (1979) Experimental manipulation of cell surface to affect cellular recognition mechanisms. *Dev. Biol.*, **70**, 195.
72. Grunwald,G.B., Pratt,R.S. and Lilien,J. (1982) Enzymic dissection of embryonic cell adhesive mechanisms. III. Immunological identification of a component of the calcium-dependent adhesive system of embryonic chick neural retina cells. *J. Cell Sci.*, **55**, 69.
73. Cook,J.H. and Lilien,J. (1981) The accessibility of certain proteins on embryonic chick neural retina cells to iodination and tryptic removal is altered by calcium. *J. Cell Sci.*, **55**, 85.
74. Hatta,K. and Takeichi,M. (1986) Expression of N-cadherin adhesion molecules associated with early morphogenetic events in chick development. *Nature*, **320**, 447.
75. Hatta,K., Takagi,S., Fujisawa,H. and Takeichi,M. (1987) Spatial and temporal expression pattern of N-cadherin cell adhesion molecules correlated with morphogenetic processes of chicken embryos. *Dev. Biol.*, **120**, 215.
76. Takeichi,M. (1988) The cadherins: cell–cell adhesion molecules controlling animal morphogenesis. *Development*, **102**, 639.

77. Hatta,K., Nose,A. Nagafuchi,A. and Takeichi,M. (1988) Cloning and expression of cDNA encoding a neural calcium-dependent cell adhesion molecule: its identity in the cadherin gene family. *J. Cell Biol.*, **106**, 873.
78. Abo,T. and Balch,C.M. (1981) A differentiation antigen of human NNK and K cells identified by a monoclonal antibody (HNK-1). *J. Immunol.*, **127**, 1024.
79. Schuller-Petrovic,S., Gebhart,W., Lassman,H., Rumpold,H. and Kraft,D. (1983) A shared antigenic determinant between natural killer cells and nervous tissue. *Nature*, **306**, 179.
80. McGarry,R.C., Helfand,S.L., Quaries,R.H. and Roder,J.C. (1983) Recognition of myelin-associated glycoprotein by the monoclonal antibody HNK-1. *Nature*, **306**, 376.
81. Chou,K.H., Ilyas,A.A., Evan,J.E., Quarles,R.H. and Jungalwala,F.B. (1985) Structure of a glycolipid reacting with monoclonal IgM in neuropathy and with HNK-1. *Biochem. Biophys. Res. Commun.*, **128**, 383.
82. Schwarting,G.A., Jungalwala,F.B., Chou,D.K.H., Boyer,A.M. and Yamamoto,M. (1987) Sulfated glucuronic acid-containing glycoconjugates are temporally and spatially regulated antigens in the developing mammalian nervous system. *Dev. Biol.*, **120**, 65.
83. Ilyas,A.A., Quarles,R.H., MacIntosh,T.D., Dobersen,M.J., Trapp,B.D., Dalakas,M.C. and Brady,R.O. (1984) IgM in a human neuropathy related to paraproteinemia binds to a carbohydrate determinant in the myelin-associated glycoprotein and to a ganglioside. *Proc. Natl. Acad. Sci. USA*, **81**, 1225.
84. Murray,N. and Steck,A.J. (1984) Indication of a possible role in a demyelinating neuropathy for an antigen shared between myelin and NK cells. *Lancet*, **1**, 711.
85. Tucker,G.C., Aoyama,H., Lipinski,M., Tursz,T. and Thiery,J.P. (1984) Identical reactivity of monoclonal antibodies HNK-1 and NC-1: conservation in vertebrates on cells derived from the neural primordium and on some leukocytes. *Cell Diff.*, **14**, 223.
86. Kruse,J., Keilhauer,G., Faissner,A., Timpl,R. and Schachner,M. (1985) The J1 glycoprotein—a novel nervous system cell adhesion molecule of the L2/HNK-1 family. *Nature*, **316**, 146.
87. Grumet,M., Hoffman,S., Crossin,K.L. and Edelman,G.M. (1985) Cytotactin, an extracellular matrix protein of neural and non-neural tissues that mediates glia – neuron interaction. *Proc. Natl. Acad. Sci. USA*, **82**, 8075.
88. Chiquet,M. and Fambrough,D.M. (1984) Chick myotendinous antigen. II. A novel extracellular glycoprotein complex consisting of large disulfide-linked subunits. *J. Cell Biol.*, **98**, 1937.
89. Hoffman,S. and Edelman,G.M. (1987) A proteoglycan with HNK-1 antigenic determinants is a neuron-associated ligand for cytotactin. *Proc. Natl. Acad. Sci. USA*, **84**, 2523.
90. Antonicek,H., Persohn,E. and Schachner,M. (1987) Biochemical and functional characterization of a novel neuron – glia adhesion molecule that is involved in neuronal migration. *J. Cell Biol.*, **104**, 1587.
91. Kornblihtt,A.R., Vibe-Pedersen,K. and Baralle,F.E. (1984) Human fibronectin: molecular cloning evidence for two mRNA species differing by an internal segment coding for a structural domain. *EMBO J.*, **3**, 221.
92. Schwarzbauer,J.E., Tamkun,J.W., Lemischka,I.R. and Hynes,R.O. (1983) Three different fibronectin mRNAs arise by alternative splicing within the coding region. *Cell*, **35**, 421.
93. Schwarzbauer,J.E., Paul,J.I. and Hynes,R.O. (1985) On the origin of species of fibronectin. *Proc. Natl. Acad. Sci. USA*, **82**, 1424.
94. Tamkun,J.W., Schwarzbauer,J.E. and Hynes,R.O. (1984) A single rat fibronectin gene generates three different mRNAs by alternative splicing of a complex exon. *Proc. Natl. Acad. Sci. USA*, **81**, 5140.
95. Yamada,K.M., Humphries,M.J., Hasegawa,T., Hasegawa,E., Olden,K., Chen,W.T. and Akiyama,S.K. (1985) Fibronectin: molecular approaches to analyzing cell interactions with the extracellular matrix. In *The Cell in Contact—Adhesions and Junctions as Morphogenetic Determinants*. G.M.Edelman and J.P.Thiery (eds), Wiley, New York, p. 303.
96. Hynes,R. (1985) Molecular biology of fibronectin. *Annu. Rev. Cell Biol.*, **1**, 67.
97. Pierschbacher,M.D., Ruoslahti,E., Sundelin,J., Lind,P. and Peterson,P.A. (1982) The cell attachment domain of fibronectin. Determination of the primary structure. *J. Biol. Chem.*, **257**, 9593.

98. Pierschbacher,M.D. and Ruoslahti,E. (1984) Cell attachment activity of fibronectin can be duplicated by small synthetic fragments of the molecule. *Nature,* **309**, 30.
99. Pierschbacher,M.D., Hayman,E.G. and Ruoslahti,E. (1981) Location of the cell-attachment site in fibronectin with monoclonal antibodies and proteolytic fragments of the molecule. *Cell,* **26**, 259.
100. Pierschbacher,M., Hayman,E.G. and Ruoslahti,E. (1983) Synthetic peptide with cell attachment activity of fibronectin. *Proc. Natl. Acad. Sci. USA,* **80**, 1224.
101. Rogers,S.L., McCarthy,J.B., Palm,S.L., Furcht,L.T. and Letourneau,P.C. (1985) Neuron specific interactions with two neurite-promoting fragments of fibronectin. *J. Neurosci.,* **5**, 369.
102. Humphries,M.J., Akiyama,S.K., Komoriya,A., Olden,K. and Yamada,K.M. (1986) Identification of an alternatively spliced site in human plasma fibronectin that mediates cell type-specific adhesion. *J. Cell Biol.,* **103**, 2637.
103. Kleinman,H.K., Cannon,F.B., Laurie,G.W., Hassell,J.R., Aumailley,M., Terranova,V.P., Martin,G.R. and DuBois-Dalcq,M. (1985) Biological activities of laminin. *J. Cell. Biochem.,* **27**, 317.
104. von der Mark,K. and Kuhl,U. (1985) Laminin and its receptor. *Biochem. Biophys.,* **823**, 147.
105. Timple,R., Rohde,H., Robey,P.G., Rennard,S.I., Foidart,S.M. and Martin,G.R. (1979) Laminin—a glycoprotein from basement membranes. *J. Biol. Chem.,* **254**, 9933.
106. Engel,J., Odermatt,E., Engel,A., Madri,J.A., Furthmayr,H., Rohde,H. and Timpl,R. (1981) Shapes, domain organizations and flexibility of laminin and fibronectin, two multifunctional proteins of the extracellular matrix. *J. Mol. Biol.,* **150**, 97.
107. Hunter,D.D., Shah,V., Merlie,J.P. and Sanes,J.R. (1989) A laminin-like adhesive protein concentrated in the synaptic cleft of the neuromuscular junction. *Nature,* **338**, 229.
108. Terranova,V.P., Rohrbach,D.H. and Martin,G.R. (1980) Role of laminin in the attachment of PAM 212 (epithelial) cells to basement membrane collagen. *Cell,* **22**, 719.
109. Dziadek,M., Paulsson,M. and Timpl,P. (1985) Identification and interaction repertoire of large forms of the basement membrane protien nidogen. *EMBO J.,* **4**, 2513.
110. Sakashita,S., Engvall,E. and Ruoslahti,E. (1980) Basement membrane glycoprotein laminin binds to heparin. *FEBS Lett.,* **116**, 243.
111. Goodman,S.L., Deutzmann,R. and von der Mark,K. (1987) Two distinct cell-binding domains in laminin can independently promote nonneural cell adhesion and spreading. *J. Cell Biol.,* **105**, 589.
112. Graf,J., Iwamoto,Y., Sasaki,M., Martin,G.R., Kleinman,H.K., Robey,F.A. and Yamada,Y. (1987) Identification of an amino acid sequence in laminin mediating cell attachment, chemotaxis, and receptor binding. *Cell,* **48**, 989.
113. Rao,C.N., Margulies,I.M.K., Tralka,T.S., Terranova,V.P., Madri,J.A. and Liotta,L.A. (1982) Isolation of a subunit of laminin and its role in molecular structure and tumor cell attachment. *J. Biol. Chem.,* **257**, 9740.
114. Terranova,V.P., Roa,C.N., Kalebic,T., Margulies,I.M. and Liotta,L.A. (1983) Laminin receptor on human breast carcinoma cells. *Proc. Natl. Acad. Sci. USA,* **80**, 444.
115. Timpl,R., Johansson,S., van Delden,V., Oberbaumer,I. and Hook,M. (1983) Characterization of protease-resistant fragments of laminin mediating attachment and spreading of rat hepatocytes. *J. Biol. Chem.,* **258**, 8922.
116. Baron-VanEvercooren,A., Kleinman,H.K., Ohno,S., Marangos,P., Schwartz,J.P. and Dubois-Dalq,M. (1982) Nerve growth factor, laminin and fibronectin promote neurite growth in human fetal sensory ganglion cultures. *J. Neurosci. Res.,* **8**, 170.
117. Edgar,D., Timpl,R. and Thoenen,H. (1984) The heparin-binding domain of laminin is responsible for its effects on neurite outgrowth and neuronal survival. *EMBO J.,* **3**, 1463.
118. Manthorpe,M., Engvall,E., Ruoslahti,E., Longo,F.M., Davis,G.E. and Varon,S. (1983) Laminin promotes neuritic regeneration from cultured peripheral and central neurons. *J. Cell Biol.,* **97**, 1882.
119. Rogers,S.L., Letourneau,P.C., Palm,S.L., McCarthy,J. and Furcht,L.T. (1983) Neurite extension by peripheral and central nervous system neurons in response to substratum-bound fibronectin and laminin. *Dev. Biol.,* **98**, 212.

120. Lindsay,R.M., Thoenen,H. and Barde,Y.-A. (1985) Placode and neural crest-derived sensory neurons are responsive at early developmental stages to brain-derived neurotrophic factor. *Dev. Biol.*, **112**, 319.
121. Davis,G.E., Manthorpe,M., Engvall,E. and Varon,S. (1985) Isolation and characterization of rat schwannoma neurite-promoting factor: evidence that the factor contains laminin. *J. Neurosci.*, **5**, 2662.
122. Lander,A.D., Fujii,D.K. and Reichardt,L.F. (1985) Laminin is associated with the 'neurite outgrowth-promoting factors' found in conditioned media. *Proc. Natl. Acad. Sci. USA*, **82**, 2183.
123. Sanes,J.R., Marshall,L.M. and McMahan,U.J. (1978) Reinnervation of muscle fiber basal lamina after removal of myofibers, differentiation of regenerating axons at original synaptic sites. *J. Cell Biol.*, **78**, 176.
124. Glicksman,M.A. and Sanes,J.R. (1983) Differentiation of motor nerve terminals formed in the absence of muscle fibres. *J. Neurocytol.*, **12**, 661.
125. Lesot,H., Kuhl,U. and von der Mark,K. (1983) Isolation of a laminin-binding protein from muscle cell membranes. *EMBO J.*, **2**, 861.
126. Malinoff,H.L. and Wicha,M.S. (1983) Isolation of a cell surface receptor protein for laminin from murine fibrosarcoma cells. *J. Cell Biol.*, **96**, 1475.
127. Rao,N.C., Barsky,S.H., Terranova,V.P. and Liotta,L.A. (1983) Isolation of a tumor cell laminin receptor. *Biochem. Biophys. Res. Commun.*, **111**, 804.
128. Wewer,U.M., Liotta,L.A., Jaye,M., Ricca,G.A., Drohan,W.N., Claysmith,A.P., Roa,C.N., Wirth,P., Coligan,J.E., Albrechtsen,R., Mudryj,M. and Sobel,M.E. (1986) Altered levels of laminin receptor mRNA in various human carcinoma cells that have different abilities to bind laminin. *Proc. Natl. Acad. Sci. USA*, **83**, 7137.
129. Pytela,R., Pierschbacher,M.D. and Ruoslahti,E. (1985) Identification and isolation of a 140 kd cell glycoprotein with properties expected of a fibronectin receptor. *Cell*, **40**, 191.
130. Brown,P.J. and Juliano,R.L. (1985) Selective inhibition of fibronectin-mediated cell adhesion by monoclonal antibodies to a cell-surface glycoprotein. *Science*, **228**, 1448.
131. Greve,J.M. and Gottlieb,D.I. (1982) Monoclonal antibodies which alter the morphology of cultured chick myogenic cells. *J. Cell. Biochem.*, **18**, 221.
132. Neff,N.T., Lowrey,C., Decker,C., Tovar,A., Damsky,C., Buck,C. and Horwitz,A.F. (1982) A monoclonal antibody detaches embryonic skeletal muscle from extracellular matrices. *J. Cell Biol.*, **95**, 654.
133. Akiyama,S.K., Yamada,S.S. and Yamada,K.M. (1986) Characterization of a 140-kd avian cell surface antigen as a fibronectin-binding molecule *J. Cell Biol.*, **102**, 442.
134. Horwitz,A., Duggan,K., Greggs,R., Decker,C. and Buck,C. (1985) The cell substrate attachment (CSAT) antigen has properties of a receptor for laminin and fibronectin. *J. Cell Biol.*, **101**, 2134.
135. Hummel,J.P. and Dryer,W.J. (1962) Measurement of protein binding phenomena by gel filtration. *Biochim. Biophys. Acta*, **63**, 530.
136. Burridge,K. and Connell,L. (1983) A new protien of adhesion plaques and ruffling membranes. *J. Cell Biol.*, **97**, 359.
137. Horwitz,A., Duggan,K., Buck,C., Beckerle,M.C. and Burridge,K. (1986) Interaction of plasma membrane fibronectin receptor with taline—a transmembrane linkaage. *Nature*, **320**, 531.
138. Tamkun,J.W., DeSimone,D.W., Fonda,D., Patel,R.S., Buck,C., Horwitz,A.F. and Hynes,R.O. (1986) Structure of integrin, glycoprotein involved in the transmembrane linkage between fibronectin and actin. *Cell*, **46**, 271.
139. Bozyczko,D. and Horwitz,A.F. (1986) The participation of a putative cell surface receptor for laminin and fibronectin in peripheral neurite extension. *J. Neurosci.*, **6**, 1241.
140. Hall,D.E., Neugebauer,K.M. and Reichardt,L.F. (1987) Embryonic neural retinal cell response to extracellular matrix proteins: developmental changes and effects of the cell substratum attachment antibody (CSAT). *J. Cell Biol.*, **104**, 623.
141. Hynes,R. (1987) Integrins: a family of cell surface receptors. *Cell*, **48**, 549.
142. Ruoslahti,E. and Pierschbacher,M.D. (1987) New perspective in cell adhesion: RGD and Integrins. *Science*, **238**, 491.
143. Suzuki,S., Oldberg,A., Hayman,E.G., Pierschbacher,M.D. and Ruoslahti,E. (1985)

Complete amino acid sequence of human vitronectin deduced for cDNA. Similarity of cell attachment sites of vitronectin and fibronectin. *EMBO J.*, **4**, 2519.
144. Bernard,M.P., Myers,J.C., Chu,M.L., Ramirez,F., Eikenberry,E.R. and Prockop,D.J. (1983) Structure of a cDNA for the Pro a2 chain of human type I procollagen. Comparison with chick cDNA for pro a2(I) identifies structurally conserved features of the protein and the gene. *Biochemistry*, **22**, 1139.
145. Doolittle,R.E., Watt,K.W.K., Cottrell,B.A., Stong,D.D. and Riley,M. (1979) The amino acid sequence of the α-chain of human fibrinogen. *Nature*, **280**, 464.
146. Poole,S., Firtel,R.A., Lamar,E. and Rowekamp,W. (1982) Sequence and expression of the discoidin I gene family in *Dictyostelium discoideum*. *J. Mol. Biol.*, **153**, 273.
147. Dedhar,S., Ruoslahti,E. and Pierrschbacher,M.D. (1987) A cell surface receptor complex for collagen type I recognizes the arg–gly–asp sequence. *J. Cell Biol.*, **104**, 585.
148. Ginsberg,M., Pierschbacher,M.D., Ruoslahti,E., Marguerie,G. and Plow,E. (1985) Inhibition of fibronectin binding to platelets by proteolytic fragments and synthetic peptides which support fibroblast adhesion. *J. Biol. Chem.*, **260**, 3931.
149. Hayman,E.G., Pierschbacher,M.D. and Rouslahti,E. (1985) Detachment of cells from culture substrate by soluble fibronectin peptides. *J. Cell Biol.*, **100**, 1948.
150. Springer,W.R., Cooper,D.N.W. and Barondes,S.H. (1984) Discoidin I is implicated in cell-substratum attachment and ordered cell migration of *Dictyostelium discoideum* and resembles fibronectin. *Cell*, **39**, 557.
151. Pytela,R., Pierschbacher,M.D. and Ruoslahti,E. (1985) A 125/115-kDa cell surface receptor specific for vitronectin interacts with the arginine-glycine-aspartic acid adhesion sequence derived from fibronectin. *Proc. Natl. Acad. Sci. USA*, **82**, 5766.
152. Pytela,R., Pierschbacher,M.D., Ginsberg,M.H., Plow,E.F. and Ruoslahti,E. (1986) Platlet membrane glycoprotein IIb/IIIa: Member of a family of Arg–Gly–Asp-specific adhesion receptors. *Science*, **231**, 1559.
153. Plow,E.F., Loftus,J.C., Levin,E.G., Fair,D.S., Dixon,D., Forsyth,J. and Ginsberg,M.H. (1986) Immunologic relationship between platelet membrane glycoprotein GPIIb/IIIa and cell surface molecules expressed by a variety of cells. *Proc. Natl. Acad. Sci. USA*, **83**, 6002.
154. Kishimoto,T.K., O'Connor,K., Lee,A., Roberts,T.M. and Springer,T.A. (1987) Cloning of the B subunit of the leukocyte adhesion proteins: homology to an extracellular matrix receptor defines a novel supergene family. *Cell*, **48**, 681.
155. Hemler,M.E., Huang,C. and Schwarz,L. (1987) The VLA protein family. Characterization of five distinct cell surface heterodimers each with a common 130,000 molecular weight b subunit. *J. Biol. Chem.*, **262**, 3300.
156. Buck,C.A., Shea,E., Duggan,K. and Horwitz,A.F. (1986) Integrin (the CSAT antigen): functionality requires oligometric integrity. *J. Cell Biol.*, **103**, 2421.
157. Fitzgerald,L.A., Steiner,B., Rall,S.C.,Jr, Lo,S.S. and Phillips,D.R. (1987) Protein sequence of endothelial glycoprotein IIIa derived from a cDNA clone. *J. Biol. Chem.*, **262**, 3936.
158. Ginsberg,M.H., Loftus,J., Ryckwaert,J.-J., Pierschbacher,M., Pytela,R., Ruoslahti,E. and Plow,E.F. (1987) Immunochemical and amino-terminal sequence comparison of two cytoadhesins indicate they contain similar or identical β subunits and distinct α subunits. *J. Biol. Chem.*, **262**, 5437.
159. Takada,Y., Huang,C. and Hemler,M.E. (1987) Fibronectin receptor structures in the VLA family of heterodimers. *Nature*, **326**, 607.
160. Suzuki,S., Argraves,W.S., Pytela,R., Arai,H., Krusius,T., Pierschbacher,M.D. and Ruoslahti,E. (1986) cDNA and amino acid sequence of the cell adhesion protein receptor recognizing vitronectin reveal a transmembrane domain and homologies with other adhesion protein receptors. *Proc. Natl. Acad. Sci. USA*, **83**, 8614.
161. Takada,Y., Strominger,J.L. and Hemler,M.E. (1987) The very late antigen family of heterodimers is part of a superfamily of molecules involved in adhesion and embryogenesis. *Proc. Natl. Acad. Sci. USA*, **84**, 3239.
162. Duband,J.L., Rocher,S., Chen,W.T., Yamada,K.M. and Thiery,J.P. (1986) Cell adhesion and migration in the early vertebrate embryo: location and possible role of the putative fibronectin receptor complex receptor complex. *J. Cell Biol.*, **102**, 160.
163. Krotoski,D.M., Domingo,C. and Bonner-Fraser,M. (1986) Distribution of a putative

cell surface receptor for fibronectin and laminin in the avian embryo. *J. Cell Biol.,* **103**, 1061.
164. Pischel,K.D., Bluestein,H.G. and Woods,V.L. (1986) Very late activation antigens (VLA) are human leukocyte–neuronal crossreactive cell surface antigens. *J. Exp. Med.,* **164**, 393.
165. Leptin,M. (1986) The fibronectin receptor family. *Nature,* **321**, 728.
166. Brower,D.L., Wilcox,M., Piovant,M., Smith,R.J. and Reger,L.A. (1984) Related cell-surface antigens expressed with positional specificity in *Drosophila* imaginal discs. *Proc. Natl. Acad. Sci. USA,* **81**, 7485.
167. Wilcox,M., Brower,D.L. and Smith,R.J. (1981) A position-specific cell surface antigen in the *Drosophilia* wing imaginal disc. *Cell,* **25**, 159.
168. Wilcox,M., Brown,N., Piovant,M., Smith,R.J. and White,R.A.H. (1984) The *Drosophila* position-specific antigens are a family of cell surface glycoprotein complexes. *EMBO J.,* **3**, 2307.
169. Wilcox,M. and Leptin,M. (1985) Tissue-specific modulation of a set of related cell surface antigens in *Drosophila*. *Nature,* **316**, 351.
170. Gallin,W.J., Edelman,G.M. and Cunningham,B.A. (1983) Characterization of L-CAM, a major cell adhesion molecule from embryonic liver cells. *Proc. Natl. Acad. Sci. USA,* **80**, 1038.
171. Ogou,S., Yoshida-Noro,C. and Takeichi,M. (1983) Calcium-dependent cell–cell adhesion molecules common to hepatocytes and teratocarcinoma stem cells. *J. Cell Biol.,* **97**, 944.
172. Thiery,J.P., Delouvee,A., Gallin,W.J., Cunningham,B.A. and Edelman,G.M. (1984) Ontogenetic expression of cell adhesion molecules: L-CAM is found in epithelia derived from the three primary germ layers. *Dev. Biol.,* **102**, 61.
173. Jacobson,M. and Rutishauser,U. (1986) Induction of neural cell adhesion molecule (NCAM) in *Xenopus* embryos. *Dev. Biol.,* **116**, 524.
174. Kintner,C.R. and Melton,D.A. (1987) Expression of *Xenopus* N-CAM RNA in ectoderm is an early response to neural induction. *Development,* **99**, 311.
175. LeDouarin,A.M. (1982) *The Neural Crest.* Cambridge University Press, Cambridge.
176. LeDouarin,N.M. (1986) Cell line segregation during peripheral nervous system ontogeny. *Science,* **231**, 1515.
177. Vincent,M., Dubard,J.L. and Thiery,J.P. (1983) A cell surface determinant expressed early in migrating avian neural crest cells. *Dev. Brain Res.,* **9**, 235.
178. Vincent,M. and Thiery,J.P. (1984) A cell surface marker for neural crest and placodal cells: further evolution in peripheral and central nervous system. *Dev. Biol.,* **103**, 468.
179. Bronner-Fraser,M. (1986) Analysis of the early stages of trunk neural crest migration in avian embryos using monoclonal antibody HNK-1. *Dev. Biol.,* **115**, 44.
180. Loring,J.F. and Erickson,C.A. (1987) Neural crest cell migratory pathways in the trunk of the chick embryo. *Dev. Biol.,* **121**, 220.
181. Rickmann,M., Fawcett,J.W. and Keynes,R.J. (1985) The migration of neural crest cells and the growth of motor axons through the rostral half of the chick somite. *J. Embryol. Exp. Morph.,* **90**, 437.
182. Tosney,K.W. (1978) The early migration of neural crest cells in the trunk region of the avian embryo: an electron microscopic study. *Dev. Biol.,* **62**, 317.
183. Tosney,K.W. (1982) The segregation and early migration of cranial neural crest cells in the avian embryo. *Dev. Biol.,* **89**, 13.
184. Mayer,B.W.,Jr, Hay,E.D. and Hynes,R.O. (1981) Immunocytochemical localization of fibronectin in embryonic chick trunk and area vasculosa. *Dev. Biol.,* **82**, 267.
185. Newgreen,D. and Thiery,J.P. (1980) Fibronectin in early avian embryos: synthesis and distribution along the migration pathways of neural crest cells. *Cell Tiss. Res.,* **211**, 269.
186. Rovasio,R.A., Delouvee,A., Yamada,K.M., Timpl,R. and Thiery,J.P. (1983) Neural crest cell migration: requirements for exogenous fibronectin and high cell density. *J. Cell Biol.,* **96**, 462.
187. Bronner-Fraser,M. (1985) Alterations in neural crest migration by a monoclonal antibody that affects cell adhesion. *J. Cell Biol.,* **101**, 610.
188. Goodman,S.I. and Newgreen,D. (1985) Do cells show an inverse locomotory response to fibronectin and laminin substrates? *EMBO J.,* **4**, 2769.

189. Bronner-Fraser,M. (1986) An antibody to a receptor for fibronectin and laminin perturbs cranial neural crest development *in vivo*. *Dev. Biol.*, **171**, 528.
190. Boucaut,J.C., Darribere,T., Poole,T.J., Aoyama,H., Yamada,K.M. and Thiery,J.P. (1984) Biologically active synthetic peptides as probes of embryonic development: a competitive peptide inhibitor of fibronectin function inhibits gastrulation in amphibian embryos and neural crest cell migration in avian embryos. *J. Cell Biol.*, **99**, 1822.
191. Tucker,G.C., Ciment,G. and Thiery,J.P. (1986) Pathways of avian neural crest cell migration in the developing gut. *Dev. Biol.*, **116**, 439.
192. Duband,J.L., Tucker,G.C., Poole,T.J., Vincent,M., Aoyama,H. and Thiery,J.P. (1985) How do the migratory and adhesive properties of the neural crest govern ganglia formation in the avian peripheral nervous system? *J. Cell. Biochem.*, **27**, 189.
193. Rakic,P. (1971) Neuron – glia relationship during granule cell migration in developing cerebellar cortex. A golgi and electron microscope study in macaque rhesus. *J. Comp. Neurol.*, **141**, 283.
194. Rakic,P. (1985) Mechanisms of neuronal migration in developing cerebellar cortex. In Edelman,G.M., Gall,W.E. and Cowan,W.M. (eds), *The Molecular Bases of Neural Development*. Wiley, New York. p. 139.
195. Chuong,C.H., Crossin,K.L. and Edelman,G.M. (1987) Sequential expression and differential function of multiple adhesion molecules during the formation of cerebellar cortical layers. *J. Cell Biol.*, **104**, 331.
196. Persohn,E. and Schachner,M. (1987) Immunoelectron microscope localization of the neural cell adhesion molecules L1 and N-CAM during postnatal development of the mouse cerebellum. *J. Cell Biol.*, **105**, 569.
197. Edmonson,J.C. and Hatten,M.E. (1987) Glial-guided granule neuron migration *in vitro*: a high-resolution time-lapse video microscopic study. *J. Neurosci.*, **7**, 1928.
198. Hatten,M.E., Liem,R.K.H. and Mason,C.A. (1984) Two forms of cerebellar glial cells interact differentially with neurons *in vitro*. *J. Cell Biol.*, **98**, 193.
199. Edmonson,J.C., Liem,R.K.H., Kuster,J.E. and Hatten,M.E. (1988) Astrotactin: a novel neuronal cell surface antigen that mediates neuron – astroglial interactions in cerebellar microcultures. *J. Cell Biol.*, **106**, 505.
200. Hatten,M.E., Liem,R.K.H. and Mason,C.A. (1986) Weaver mouse cerebellar granule neurons fail to migrate on wild-type astroglial processes *in vitro*. *J. Neurosci.*, **6**, 2676.
201. Lindner,J., Rathjen,F.G. and Schachner,M. (1983) L1 mono- and polyclonal antibodies modify cell migration in eary postnatal mouse cerebellum. *Nature*, **305**, 427.
202. Hoffman,S., Friedlander,D.R., Chuong,C., Grumet,M. and Edelman,G.M. (1986) Differential contributions of Ng-CAM and N-CAM to cell adhesion in different neural regions. *J. Cell Biol.*, **103**, 145.
203. Holt,C. and Harris,W.A. (1983) Order in the initial retinotectal map in *Xenopus*: a new technique for labelling growing nerve fibers. *Nature*, **301**, 150.
204. Tosney,K.W. and Landmesser,L.T. (1985) Specificity of early motoneuron growth cone outgrowth in the chick embryo. *J. Neurosci.*, **5**, 2336.
205. Lockerbie,R.O. (1987) The neuronal growth cone: a review of its locomotory, navigational and target recognition capabilities. *Neuroscience*, **20**, 719.
206. Letourneau,P.C. (1979) Cell – substratum adhesion of neuritic growth cones, and its role in neurite elongation. *Exp. Cell Res.*, **124**, 127.
207. Letourneau,P.C. (1975) Cell-to-substratum adhesion and guidance of axonal elongation. *Dev. Biol.*, **44**, 92.
208. Gundersen,R.W. (1987) Response of sensory neurites and growth cones to patterned substrata of laminin and fibronectin *in vitro*. *Dev. Biol.*, **121**, 423.
209. Bray,D. (1984) Axonal growth in response to experimentally applied mechanical tension. *Dev. Biol.*, **102**, 379.
210. Caudy,M. and Bentley,D. (1986) Pioneer growth cone morphologies reveal proximal increases in substrate affinity within leg segments of grasshopper embryos. *J. Neurosci.*, **6**, 364.
211. Nardi,J.B. (1983) Neuronal pathfinding in developing wings of the moth *Manduca sexta*. *Dev. Biol.*, **95**, 163.
212. Akers,R.M., Mosher,D.F. and Lilien,J.E. (1981) Promotion of retinal neurite outgrowth by substratum-bound fibronectin. *Dev. Biol.*, **86**, 179.
213. Collins,F. (1978) Induction of neurite outgrowth by a conditioned medium factor bound to culture substratum. *Proc. Natl. Acad. Sci. USA*, **75**, 5210.

214. Chiu,A.Y., Matthew,W.D. and Patterson,P.H. (1986) A monoclonal antibody that blocks the activity of a neurite regeneration-promoting factor: studies on the binding site and its localization in vivo. *J. Cell Biol.*, **102**, 1383.
215. Matthew,W.D. and Patterson,P.H. (1983) The production of a monoclonal antibody that blocks the action of a neurite outgrowth-promoting factor. *Cold Spring Harbor Symp. Quant. Biol.*, **48**, 625.
216. Lander,A.D., Fudjii,D.K. and Reichardt,L.F. (1985) Purification of a factor that promotes neurite outgrowth: isolation of laminin and associated molecules. *J. Cell Biol.*, **101**, 898.
217. Hantaz-Ambroise,D., Vigny,M. and Koenig,J. (1987) Heparan sulfate proteoglycan and laminin mediate two different types of neurite outgrowth. *J. Neurosci.*, **7**, 2293.
218. Rogers,S.L., Edson,K.J., Letourneau,P.C. and McLoon,S.C. (1986) Distribution of laminin in the developing peripheral nervous system of the chick. *Dev. Biol.*, **113**, 429.
219. Sandrock,A.W.,Jr, and Matthew,W.D. (1987) An *in vitro* neurite-promoting antigen functions in axonal regeneration *in vivo*. *Science*, **237**, 1605.
220. Laverty,R., Sharman,D.F. and Vogt,M. (1965) Action of 2,3,5-trihydroxyphenylethylamine on storage and release of noradrenaline. *Br. J. Pharmacol. Chemother*, **24**, 549.
221. Cohen,J., Burne,J.F., McKinlay,C. and Winter,J. (1987) The role of laminin/fibronectin receptor complex in the outgrowth of retinal ganglion cell axons. *Dev. Biol.*, **122**, 407.
222. Cohen,J., Burne,J.F., Winter,J. and Bartlett,P. (1986) Retinal ganglion cells lose response to laminin with maturation. *Nature*, **322**, 465.
223. Liesi,P. (1985) Laminin-immunoreactive glia distinguish regenerative adult CNS systems from non-regenerative ones. *EMBO J.*, **4**, 2505.
224. Carbonetto,S., Evans,D. and Cochard,P. (1987) Nerve fiber growth in culture on tissue substrata from central and peripheral nervous systems. *J. Neurosci.*, **7**, 610.
225. Rutishauser,R., Gall,W.E. and Edelman,G.M. (1978) Adhesion among neural cells of the chick embryo IV. Role of the cell surface molecule CAM in the formation of neurite bundles in cultures of spinal ganglia. *J. Cell Biol.*, **79**, 382.
226. Chang,S., Rathjen,F.G. and Raper,J.A. (1987) Extension of neurites on axons is impaired by antibodies against specific neural cell surface glycoproteins. *J. Cell Biol.*, **104**, 355.
227. Rutishauser,R., Watanabe,M., Silver,J., Troy,F.A. and Vimr,E.R. (1985) Specific alteration of NCAM-mediated cell adhesion by an endoneuraminidase. *J. Cell Biol.*, **101**, 1842.
228. Stallcup,W.B. and Beasley,L. (1985) Involvement of the nerve growth factor-inducible large external glycoprotein (NILE) in neurite fasciculation in primary cultures of rat brain. *Proc. Natl. Acad. Sci. USA*, **82**, 1276.
229. Beasley,L. and Stallcup,W.B. (1987) The nerve growth factor-inducible large external (NILE) glycoprotein and neural cell adhesion molecule (N-CAM) have distinct patterns of expression in the developing rat central nervous system. *J. Neurosci.*, **7**, 708.
230. Daniloff,J.K., Chuong,C., Levi,G. and Edelman,G.M. (1986) Differential distribution of cell adhesion molecules during histogenesis of the chick nervous system. *J. Neurosci.*, **6**, 739.
231. Stallcup,W.B., Beasley,L.L. and Levine,J.M. (1985) Antibody against nerve growth factor-inducible large external (NILE) glycoprotein labels nerve fiber tracts in the developing rat nervous system. *J. Neurosci.*, **5**, 1090.
232. Thiery,J.P., Delouvee,A., Grumet,M. and Edelman,G.M. (1985) Initial appearance and regional distribution of the neuron–glia cell adhesion molecule in the chick embryo. *J. Cell Biol.*, **100**, 442.
233. Rathjen,F.G., Wolff,J.M., Frank,R., Bonhoeffer,F. and Rutishauser,U. (1987) Membrane glycoproteins involved in neurite fasciculation. *J. Cell Biol.*, **104**, 343.
234. Vimr,E.R., McCoy,R.D., Vollger,H.F., Wilkinson,N.C. and Troy,F.A. (1984) Use of prokaryotic-derived probes to identify poly(sialic acid) in neonatal neuronal membranes. *Proc. Natl. Acad. Sci. USA*, **81**, 1971.
235. Rutishauser,U., Acheson,A., Hall,A.K., Mann,D.M. and Sunshine,J. (1988) The neural cell adhesion molecule (NCAM) as a regulator of cell–cell interactions. *Science*, **240**, 53.
236. Sperry,R.W. (1963) Chemoaffinity in the orderly growth of nerve fiber patterns and connections. *Proc. Natl. Acad. Sci. USA*, **50**, 703.
237. Trisler,D. and Collins,F. (1987) Corresponding spatial gradients of TOP molecules

in the developing retina and optic tectum. *Science*, **237**, 1208.
238. Fraser,S.E. (1985) Cell interactions involved in neuronal patterning: An experimental and theoretical approach. In *The Molecular Bases of Neural Development*. G.M.Edelman, W.E.Gall and W.M.Cowan (eds), Wiley, New York, p. 481.
239. Fraser,S.E. (1980) A differential adhesion approach to the patterning of nerve connections. *Dev. Biol.*, **79**, 453.
240. Whitelaw,V.A. and Cowan,J.D. (1981) Specificity and plasticity of retinotectal connections: A computational model. *J. Neurosci.*, **1**, 1369.
241. Thanos,S., Bonhoeffer,F. and Rutishauser,U. (1984) Fiber–fiber interaction and tectal cues influence the development of the chicken retinotectal projection. *Proc. Natl. Acad. Sci. USA*, **81**, 1906.
242. Fraser,S.E., Murray,B.A., Chuong,C. and Edelman,G.M. (1984) Alteration of the retinotectal map in *Xenopus* by antibodies to neural cell adhesion molecules. *Proc. Natl. Acad. Sci. USA*, **81**, 4222.
243. Fraser,S.E. and Bayer,V. (1985) An assay of cell interactions that play a role in neuronal patterning. In *The Cell in Contact: Adhesions and Junctions as Morphogenetic Determinants*. G.M.Edelman and J.P.Thiery (eds), Wiley, New York, p. 93.
244. Covault,J. and Sanes,J.R. (1986) Distribution of N-CAM in synaptic and extrasynaptic protions of developing and adult skeletal muscle. *J. Cell Biol.*, **102**, 716.
245. Edelman,G.M., Gallin,W.J., Delouvee,A., Cunningham,B.A. and Thiery,J.P. (1983) Early epochal maps of two different cell adhesion molecules. *Proc. Natl. Acad. Sci. USA*, **80**, 4384.
246. Tosney,K.W., Watanabe,M., Landmesser,L. and Rutishauser,U. (1986) The distribution of NCAM in the chick hindlimb during axon outgrowth and synaptogenesis. *Dev. Biol.*, **114**, 437.
247. Rutishauser,R., Grumet,M. and Edelman,G.M. (1983) Neural cell adhesion molecule mediates initial interactions between spinal cord neurons and muscle cells in culture. *J. Cell Biol.*, **97**, 145.
248. Bixby,J.L., Pratt,R.S., Lilien,J. and Reichardt,L.F. (1987) Neurite outgrowth on muscle cell surfaces involves extracellular matrix receptors as well as Ca-dependent and -independent cell adhesion molecules. *Proc. Natl. Acad. Sci. USA*, **84**, 2555.
249. Landmesser,L., Dahm,L., Schultz,K. and Rutishauser,U. (1988) Distinct roles for adhesion molecules during innervation of embryonic chick muscle. *Dev. Biol.*, **130**, 645.
250. Sanes,J.R., Schachner,M. and Covault,J. (1986) Expression of several adhesive macromolecules (N-CAM, L1, J1, NILE, Uvomorulin, Laminin, Fibronectin, and a Heparan sulfate proteoglycan) in embryonic, adult, and denervated adult skeletal muscle. *J. Cell Biol.*, **102**, 420.
251. Moore,S.E. and Walsh,F.S. (1985) Specific regulation of N-CAM/D2-CAM cell adhesion molecule during skeletal muscle development. *EMBO J.*, **4**, 623.
252. Rieger,F., Grumet,M. and Edelman,G.M. (1985) N-CAM at the vertebrate neuromuscular junction. *J. Cell Biol.*, **101**, 285.
253. Covault,J. and Sanes,J.R. (1985) Neural cell adhesion molecule (N-CAM) accumulates in denervated and paralyzed skeletal muscles. *Proc. Natl. Acad. Sci. USA*, **82**, 4544.
254. Moore,S.E. and Walsh,F.S. (1986) Nerve dependent regulation of neural cell adhesion molecule expression in skeletal muscle. *Neuroscience*, **18**, 499.
255. Sanes,J.R. and Covault,J. (1985) Axon guidance during reinnervation of skeletal muscle. *Trends Neurosci.*, **8**, 523.
256. Dickson,G., Gower,H.J., Barton,C.H., Prentice,H.M., Elsom,V.L., Moore,S.E., Cox,R.D., Quinn,C., Putt,W. and Walsh,F.S. (1987) Human muscle neural cell adhesion molecule (N-CAM): identification of a muscle-specific sequence in the extracellular domain. *Cell*, **50**, 1119.
257. Kapfhammer,J.P. and Raper,J.A. (1987) Collapse of growth cone structure on contact with specific neurites in culture. *J. Neurosci.*, **7**, 201.
258. Kapfhammer,J.P. and Raper,J.A. (1987) Interactions between growth cones and neurites growing from different neural tissues in culture. *J. Neurosci.*, **7**, 1595.

Index

Ace, acetylcholinesterase gene of *Drosophila*, 20
acetylcholine, 5, 18
acetylcholine receptor, 63, 64
  autoimmune disease and, 64
  amino acid sequence of, 76, 77
  binding site of, 80
  bovine, 88
  BuTx binding and, 103
  cDNA in retroviral vectors and, 102
  cDNAs of, 76
  channel-lining sequences of, 93–94
  chimeric receptors, 104
  chromosomal genes for, 112
  cytoplasmic location of C-terminus in, 82
  electron image analysis of crystals of, 68
  expression of in cultured cell lines, 100
  expression of in *Escherichia coli*, 98
  expression of in fibroblast cells, 102
  expression of in *Xenopus* oocytes, 99
  expression of in yeast, 100
  glycosylation sites, site-directed mutagenesis of, 98
  human, 88
  isolation of novel subunits in, 95
  levels of subunit mRNA and denervation in, 113
  membrane-spanning domains in, 82, 83, 93
  murine, 88
  neuronal, 114
  of chicken, 88
  of *Drosophila*, 18, 88, 115, 118
  of elasmobranch *Torpedo*, 67
  of *Electrophorus*, 67
  of rat, 88
  of *Torpedo*, 67
  of *Xenopus*, 88, 97, 114
  order of subunit assembly in, 109
  post-translational modifications in, 68
  potential glycosylation sites in, 77
  potential phosphorylation sites in, 79
  signal peptide of, 76
  site-directed mutagenesis in, 85, 98
  subunits of, 68, 76
  synaptogenesis in, 64
  topology of subunit folding in, 79, 81
  Torpedo–mouse hybrids of, 106
  Torpedo–calf hybrids of, 104
  Torpedo–mouse hybrids of, 104
  transcripts of in *Xenopus* development, 114
  transmembrane domains in, 82, 83, 93
  X-ray crystallographic studies of, 86
acetylcholinesterase (AChE), 17, 18
actinin, 71
adenosine, 5
adhesion molecule on glia (AMOG), 160
adrenergic receptor, 5
agrin,
  150 kd, 71
  95 kd, 71
alkaloid toxins, 69
AMOG, *see* adhesion molecule on glia
anticonvulsants (barbiturates), 65
anti-NCAM Fabs and defasciculation, 177
anxiolytics (benzodiazepines), 65
astrocytic processes, 172
astrotactin, 144, 172
atrial natriuretic peptide, 5
autoimmune disease myasthenia gravis, 74
axon,
  fasciculation of, 146, 175
  growth of, 175
axon–axon interactions, 177

basal lamina, 71, 145, 162
Bergmann glia, 144, 170
  fibres of, 170
biotinylated cobra toxin, 108
bungarotoxin, 18, 108

calcium channels, 65, 66, 122
  in muscle, 66
calcium-dependent adhesion, 144
calcium-dependent neural adhesion

molecules, 157
cAMP-dependent protein kinase, 80
cell surface receptor for laminin and fibronectin, 145
cerebellar granule cells, 144
  migration of, 170
cervical giant fibers, 12
chick optic nerve, 176
chlorpromazine, 69
choline acetyltransferase, 17, 18
chromosome jumping, 16
chromosome walk,
  through *Ace* region in *Drosophila*, 21
  through *Sh* region in *Drosophila*, 47
collagen type IV, 162
convulsants (picrotoxin), 65
curare, rate of AChR synthesis and, 71
curaremimetic neurotoxins, 68, 69
cytoadhesion family, 165, 188
cytotactin, 160

development of neural tube, 167
dihydropyridine-sensitive calcium channel, 66
dimethyl-*d*-tubocurarine, hexamethonium, 69
discoidin 1, an adhesion protein of *Dictyostelium*, 165
dopamine, 5
dorsal longitudinal flight muscles (DLMs), 13
double mutants in *Drosophila*, 41
  Sh and *eag*, 46, 50
  para and *tip-E*, 42
Drosophila homolog of the vertebrate Na$^+$ channel, 31
Drosophila position-specific antigens, 166

eag (ether a-go-go) locus of *Drosophila*, 16, 46, 49
  leg-shaking behavior in, 49
electrocytes from *Torpedo*, 67
Electrophorus, 67
electrophysiological abnormalities in the GFP of *Drosophila*, 19
electrophysiology of AChR function, 70
electroretinogram, 13
  defective mutants, 30
entactin/nidogen, 162
external cell layer, 170
extracellular matrix adhesion molecules, 161

fasciculation of growing retinal axons, 183
fate of crest cells quail-chick, 168
fibronectin, 145, 161
filamin, 71
fodrin and NCAM, 150

GABA, 5

GABA receptor, 65, 117, 118
  G proteins and, 65
  linked to chloride channel, 65
germ line transformation of *Drosophila*, 18
GFP, *see* giant fiber pathway
giant fiber pathway (GFP), 13
glutamate, 5
glycine receptor, 5, 65, 117–119
  strychnine antagonist for, 65
  strychnine binding subunit, similarity of to AChR subunits, 119
granule cell layer, 170
granule cell migration, 171
  effects of anti-NCAM antibodies on, 173, 174
granule cell–astroglial interactions, 171
growth cone, 175
growth-factor-type neuropeptide, 5
guanylate cyclase peptide receptors, 5
G-protein-dependent second messenger system, 105

hallucinogens, 69
heparan sulfate proteoglycan, 162
histamine, 5
histrionicotoxin, 69
Hk (Hyperkinetic) leg-shaking, 131
HNK-1 epitope, 145
hybrid dysgenesis, 16

immunoglobulin superfamily, 154
integrins, 145, 164, 165
integrin CSAT/JG22 antigen,
  binding of to fibronectin, 164
  binding of to talin, 164

J1/cytotactin, 144, 158, 159, 171

kainate-type, 5

L1 NgCAM, 146, 171
  on post-mitotic neurons, 152
L2/HNK-1 J1, 159
laminin, 145, 162
  as a neurite outgrowth-promoting factor, 175, 176
  fibronectin and, cell surface receptors for, 163
  inducer of neurite outgrowth, 162
laminin–CSAT integrin, 164
LCAM during neural induction, 167
lidocaine, 69

MAG, (*see* myelin-associated glycoprotein)
medioventral longitudinal muscle of larvae, 12
N-methyl-D-aspartate-type, 5
migration of crest cells, 169
molecular layer, 170

monoclonal antibodies,
  JG22, 164
  CSAT, 164
  15G8, 150
mutations affecting sodium channels, 34
myelin-associated glycoproteins, 146, 153, 154

nap locus of Drosophila, 16, 34–36
NCAM, 72, 146, 147, 171
  AMOG and, 171
  N-cadherin and, expression of by embryonic muscle, 184
  during neural induction, 167
  expression of in the formation of ganglia, 170
  fasciculation and, 179
  homophilic binding and, 149
  of myotubes, 185
  on glial cells, 151
nerve growth factor, 5
nerve–muscle adhesion, 184
neural crest cell migration, 168
neural induction, NCAM and LCAM during, 167
neurite outgrowth along axon, 180
neuroleptics, 69
neuron–glial cell adhesion, 152
neurotrophic factors, 71
NgCAM, 152
ninaE locus of Drosophila, 25
NMDA-type, see N-methyl-D-aspartate-type
norpA (no receptor potential), locus, 29, 30
N-cadherin, 144, 157, 158
  LCAM and, 158
  uvomorulin and, 158

oligodendrocyte proteolipid protein, 153
oligodendroglia, 146
ommatidium, 25
oocyte expression system, 74, 99

P elements, 16
para locus of Drosophila, 35–37
  transposon tagging and, 37
phencyclidine, 69
pheochromacytoma cell line PC12, 115, 153
phospholipase-C, 30
phototransduction, 23
polysialic acid on NCAM, 148
poly[A]$^+$ mRNA from Torpedo marmorata, 74
potassium channel, 33, 35, 43, 50, 65, 66, 67
  Drosophila cDNA, 67
  mutants, 35
  see also Sh
protein kinase C (PKC), 80
proteoglycan binding, 150
Purkinje cell layers, 170

quantum release of ACh per vesicle, 64
quinacrine, 69

R1–R6 opsin gene: ninaE, 26
R7 opsin genes, 28
R8 opsin gene, 27
rhabdomers, 25

saxitoxin (STX), 35
Schwann cell, 144, 146, 187
second messengers in invertebrate phototransduction, 30
segmental aneuploidy, 17
sei locus of Drosophila, 39
  STX-binding of, 40
Sh (Shaker locus of Drosophila), 16, 33, 44–46, 123
  electrophysiological defects and, 49
sialic acid associated with NCAM, 150
slo (slowpoke), 52
sodium channels, 15, 31, 36, 48, 66
spectrin, 71
STX, see saxitoxin
sulfated glucuronic acid, 145
superfamilies of ligand-gated ion channels, 5
synaptic cleft of nerve–muscle synapses, 163
S-laminin, 145, 163

talin, 71
tetrodotoxin (TTX), 34
tip-E (temperature-induced paralysis E) locus of Drosophila, 39
transient receptor potential (trp), 29
transposon tagging, 16
trimethisoquin, 69
TTX, see tetrodotoxin
tyrosine kinase, 80

vinculin, 71

weaver, 172